*f***P**

ALSO BY DAVID LINDLEY

The End of Physics

Where Does the Weirdness Go?

The Science of Jurassic Park (with Rob DeSalle)

BOLTZMANN'S ATOM

THE GREAT DEBATE THAT LAUNCHED A REVOLUTION IN PHYSICS

David Lindley

The Free Press
New York London Toronto Sydney Singapore

THE FREE PRESS
A Division of Simon & Schuster, Inc.
1230 Avenue of the Americas
New York, NY 10020

Designed by Gabriel Levine

Manufactured in the United States of America

10 9 8 7 6 5 4 3 2 1

Library of Congress Cataloging-in-Publication Data Is Available

Lindley, David
Boltzman's atom : the great debate that launched a revolution
in physics / David Lindley.
p. cm.
Includes bibliographical references and index.
1. Boltzmann, Ludwig, 1844–1906. 2. Physicists–Austria–Biography.
3. Atomic theory–History–19th century. I. Title.

QC16.B64 L56 2001
530'.092–dc21
[B] 00-32167
ISBN 978-1-5011-4244-4

Contents

Introduction

"I DON'T BELIEVE THAT ATOMS EXIST!"

As recently as the waning years of the 19th century, a respected physicist and philosopher could make that statement before an audience of his colleagues and expect to be met not with derision or ridicule but with thoughtful consideration. Many scientists at the time subscribed to the idea that all matter was made up of tiny, elemental particles called atoms, but their arguments were as yet circumstantial. No one could say exactly what an atom was. For such reasons, critics maintained that atoms were no more than clever speculation, unworthy of scientific consideration.

This blunt declaration of disbelief came, in fact, in January 1897 at a meeting of the Imperial Academy of Sciences in Vienna. The skeptic was Ernst Mach, not quite 60 years old, who had been for many years a professor of physics at the University of Prague and who was now a professor of the history and philosophy of science in Vienna. He pronounced his uncompromising opinion in the discussion following a lecture delivered by Ludwig Boltzmann, a theoretical physicist. Boltzmann, a few years younger than Mach, had likewise recently returned to Vienna after many years at other universities in Austria and Germany. He was an unabashed believer in the atomic hypothesis–indeed, his life's work had centered on that single theme.

Nowadays atoms are uncontroversial. Scientists have proved not only that they exist, but that they are made of still smaller objects.

An atom is a cloud of electrons swirling around a dense nucleus; the nucleus contains protons and neutrons; inside them are quarks. Quarks, in all likelihood, are not truly fundamental particles but manifestations of some still more elementary theoretical structure.

It seems today unremarkable, indeed a matter of course, that theoretical physicists striving to understand the fundamental nature of the world should deal in esoteric ideas and strange objects far removed from the world around us. But it was not always so. Well into the latter half of the 19th century, most scientists saw their essential task as the measurement and codification of phenomena they could investigate directly: the passage of sound waves through air, the expansion of gas when heated, the conversion of heat to motive power in a steam engine. A scientific law was a quantitative relationship between one observable phenomenon and another.

But there came a time when, to understand more deeply, a few scientists found themselves impelled to dig deeper, to probe beyond surface appearances. Ludwig Boltzmann was such a pioneer. He understood before most of his contemporaries that if he pictured a gas as a lively collection of atoms, he could explain many of its properties. The atoms' incessant motion would produce the properties recognized as temperature and pressure. Instead of merely observing and recording that a heated gas expands, he could say why it would expand, and by how much. By understanding the behavior of atoms, he could understand the ability of hot gas to push on a piston, as in a steam engine, and turn its energy into mechanical work.

In his lifelong pursuit of this new atomic perspective, moreover, Boltzmann introduced wholly new theoretical concepts into physics. Because atoms are so numerous, and their motion so varied, he had to use techniques of statistics and probability to depict their collective activities. Although atoms move in fundamentally random ways, Boltzmann found he could nonetheless make accurate predictions of their collective effects; he proved that the disorderly actions of individual atoms could give rise to orderly behavior in bulk. He showed that laws of physics could be built on a foundation of probability, and yet still be reliable. To an audience of physicists raised in the belief that scientific laws ought to encapsulate absolute certainties and unerring rules, these were profound and disturbing changes.

In 19th-century Europe, not every scientist saw such novelties as developments to be cheered. Boltzmann encountered opposition. Many of his fellow physicists did not believe that his goals were worthwhile, or even qualified as science. They had measured the expansion of gases and could write down a simple law relating temperature, pressure, and volume. Boltzmann's alleged atoms, by contrast, were invisible, intangible, and imperceptible. What was the point of explaining a straightforward law, derived directly from experiment, in terms of hypothetical entities that could not be seen and might never be seen? This was why Ernst Mach proclaimed that he didn't believe that atoms existed.

Belief, in the context of a scientific debate, may seem an odd word. Is not science a matter of proof and reason, logic and fact? Atoms, surely, either exist or they do not. Where does belief come into it?

Nevertheless, Mach used that word particularly, and he meant it. Scientific certainty is achieved only gradually–some would say never. When ideas are new and theories tentative, scientists do not and cannot have proof that they are on the right track. They generate hypotheses (a fancy word for guesses) and try to follow where their insight and imagination lead them. But rarely does a scientific hypothesis–at least, not a useful or an interesting one–admit of a straightforward up-or-down, yes-or-no verdict. Valuable hypotheses survive the test of time in countless engagements with reality. The war against ignorance is a war of attrition.

In the meantime, scientists have to keep the faith. They pursue their chosen hypotheses because they believe they are moving forward.

Boltzmann's pursuit of the atomic hypothesis sprang from just such a belief. By explaining a wide variety of the properties of gases from a single starting point, he believed he was providing a new and powerful form of understanding. Mach thought otherwise. He did not doubt Boltzmann's acuity of mind or ingenuity with theories. He just didn't see the point of all that theorizing. And he evolved a philosophy to bolster his beliefs. Science ought to stick to what it can measure directly, and theories ought to restrict themselves to specifying exact relationships between those measured phenomena. Put some energy into a gas, heat it up, and it expands. The rules for such changes had been found out and established years before. Nothing further need be said.

The debate between Boltzmann and Mach was, therefore, less about atoms themselves and more about the purpose of doing physics, and about the nature of the understanding or explaining that physicists sought to achieve. Mach argued for sticking to simple equations linking tangible quantities. Boltzmann believed that more elaborate explanations, dependent on larger assumptions or hypotheses, nevertheless provided a more complete, more satisfying view of the physical world. At the cost of introducing theoretical ideas, Boltzmann claimed, he could generate a more valuable understanding of the way the world worked.

But value, like belief, is another word that doesn't quite sound scientific. Mach, through the course of his long and productive career, steered from physics into philosophy precisely because he grew fascinated by the question of value. What is the worth of a scientific explanation? What kinds of explanations should scientists aim for? By the time he returned to Vienna, his reputation rested on his philosophical writings more than on his scientific achievements (which were diverse and useful, but none of them truly remarkable. If his name is known to nonscientists today, it is because of the Mach number, the velocity of a projectile as a multiple of the speed of sound. A convenient notion, but hardly the product of genius).

Boltzmann, on the other hand, had the impatience with philosophical quibbling typical of most scientists. As his theories grew in power and scope, he knew he was making progress. He understood things better. He didn't worry about what he meant by saying that he understood things better.

When, in 1897, Mach stood up in the Viennese Academy of Sciences and declared flatly that he didn't believe in the existence of atoms, his words, as Boltzmann recalled later, "ran around in my head." Mach's objection set him thinking–thinking, specifically, that as a theoretical physicist of undoubted prowess he ought to be able to summon up some sort of argument, of a philosophical nature, that would dent Mach's stubborn skepticism. He had never been very interested in philosophy, but he would learn something about it and refute his critics.

It was to be an unhappy impulse.

CHAPTER 1

A Letter from Bombay

Lessons in Obscurity

ON DECEMBER 11, 1845, A LENGTHY MANUSCRIPT arrived in the London offices of the Royal Society, the highest scientific association in Great Britain. The author of this work hoped his essay might be published in the Society's august *Philosophical Transactions,* and the manuscript, by standard practice, was duly sent to a couple of experts for evaluation of its worth. "Nothing but nonsense" was the verdict of one of these eminent reviewers. The other allowed that the paper demonstrated "much skill and many remarkable accordances with the general facts," but concluded nevertheless that the ideas were "entirely hypothetical" and, in the end, "very difficult to admit."

On these recommendations, the manuscript was never published. Worse still, the author, one John James Waterston, never found out what had happened. Waterston was living at the time in Bombay, teaching navigation and gunnery to naval cadets employed by the East India Company. Born and educated in Edinburgh, he spent his life working as a civil engineer and teacher, retiring from his position in India in 1857 to return to Scotland, where he lived modestly on his savings and continued to dabble in science: astronomy, chemistry, and physics. He was known during his lifetime, if at all, as one of the numerous amateurs of Victorian science, working in isolation, contributing from time to time ideas that were more or less sound but of no great consequence.

His rejected manuscript of 1845 embodied Waterston's one truly innovative and profound piece of work, but it was ahead of its time. Only by a few years, admittedly, but that was enough to ensure its unhappy reception by the experts of the Royal Society. Waterston proposed that any gas consisted of numerous tiny particles–he called them molecules–bouncing around and colliding with each other. He showed that the energy of motion in these particles corresponded to the temperature of the gas, and that the incessant impacts of the particles on the walls of the container gave rise to the effect commonly known as pressure. There was more: Waterston calculated the "elasticity" of gases (their ability to flow, roughly speaking) from his model, and he made the subtle observation that in a mixture of different gases all the tiny particles would, on average, have the same energy, so that heavier molecules would move more slowly than lighter ones. He was not right in every detail, but his general arguments and suppositions have survived the test of time. Waterston's fundamental idea, that a gas is made of tiny, colliding particles whose microscopic behavior produces the measurable properties of the gas as a whole, was exactly right.

Waterston's calculations were somewhat rough and ready, and his proofs were not quite solid. It may have been these deficiencies that led to the rejection of his paper–that, and the fact that his name was unknown. It was certainly not revolutionary, in the middle of the 19th century, to suggest that gases consisted of tiny particles. The terms *atom* and *molecule* were known in scientific circles, although they designated objects whose true nature was unclear. Even the idea that the motion and collision of these particles had something to do with temperature and pressure was not altogether new. The Royal Society, admirably consistent, had in fact rejected a very similar proposal some 25 years earlier. The author of this earlier attempt was John Herapath, another unsung amateur of Victorian science and engineering. His work was by no means as sophisticated as Waterston's, but he had the right general idea: heat equals the motion of atoms or molecules. He wrote up his ideas in 1820 and sent them to the Royal Society. The chemist Humphrey Davy, then president of the society, declined to publish the paper. Though he was not unsympathetic to atomic thinking, Davy found Herapath's calculations unconvincing, and in truth, Herapath was confused about the mechanics of atomic collisions and came up with an incorrect formula for the temperature of a gas. Still, Herapath succeeded in getting accounts of his work published in other scientific journals, where they were roundly ignored by the scientific community of that age.

Waterston knew of Herapath's work, and of his erroneous formula for temperature, but neither of these two men, it appears, was aware that the atomic picture of a gas was close to a century old by the time they came to it. In 1738 Daniel Bernoulli, one of an extended Swiss clan of Bernoullis that made notable contributions to both mathematics and physics, succeeded in deriving theoretically a relationship between the pressure exerted by a gas and the energy of vibration of the supposed atoms within it. His theory attracted little attention, and was soon forgotten.

Bernoulli's was the first modern atomic or molecular model of a gas. He explained pressure in terms of atomic motion, but not temperature, largely because the nature of heat itself was quite mysterious in Bernoulli's day. Even so, neither he nor Herapath nor Waterston can take any credit for the idea of atoms themselves. They were the inheritors of a centuries-old tradition in natural philosophy according to which everything in the universe is composed fundamentally of minute, indivisible objects. The word *atom* is of Greek origin, meaning "uncuttable," and it is from ancient Greece that the idea itself descends.

Knowledge of the atomic hypothesis from ancient times is owed largely to the survival of a long poem called *De Rerum Natura* (*On the Nature of Things*) by the Roman writer Lucretius. The names of both this poem and its author had faded into oblivion in the centuries after the fall of Rome, but a church official traveling around the monasteries of France and Germany in the 13th century happened across a copy (not an original) and brought it back to the Vatican in 1417. Manuscripts dating back to the 9th or 10th century were subsequently rediscovered and found to be substantially the same as the Vatican copy. From these versions descend all modern editions of *De Rerum Natura*. Its author, Titus Lucretius Carus, lived from about 95 to 55 B.C. The six books of his great opus lay out a philosophical reflection on life as well as an exposition of a scientific hypothesis. It is fiercely atheistic. It enjoyed a good deal of renown in its time, but was later attacked by the Emperor Augustus in his attempt to restore some of the faded glory of the declining Roman world by reviving the ancient pre-Christian religion.

Lucretius derived his atheism from his adherence to what can be called, with the benefit of two millennia of hindsight, an atomic theory of the natural world. For example:

> clothes hung above a surf-swept shore
> grow damp; spread in the sun they dry again.
> Yet it is not apparent to us how
> the moisture clings to the cloth, or flees the heat.
> Water, then, is dispersed in particles,
> atoms too small to be observable.

In other words, a wet garment has atoms (we would now say molecules) of water clinging to its fabric; heat drives the atoms off, and thus dries the material. An atomic theory of clothes drying seems to be some way from disproving the existence of deities, but Lucretius goes on to observe that the atoms have no volition, and instead move willy-nilly:

> For surely the atoms did not hold council, assigning
> order to each, flexing their keen minds with
> questions of place and motion and who goes where.

> But shuffled and jumbled in many ways, in the course
> of endless time they are buffeted, driven along,
> chancing upon all motions, combinations.
> At last they fall into such an arrangement
> as would create this universe . . .

Examined closely, Lucretius says, the range and variety of all the familiar phenomena of the world about us arise from invisible atoms zipping aimlessly this way and that. No need for gods to direct events, or inspire actions and consequences. On the other hand, Lucretius's vision seems to leave little room for human decision or free will either. If the universe takes its course because atoms are following their random paths, then neither gods nor human beings have any control over their destinies; what will happen, will happen, and there is nothing anyone can do to change it.

This is a bleak form of atheism, implying what is nowadays called determinism, meaning that what happens in the future is wholly determined by what has happened in the past. To Lucretius and his followers this view was nevertheless a liberation. In their day the gods were fickle, cruel, and capricious, more inclined to pranks and practical jokes than to love or compassion. The citizens of Rome decidedly did not wish for a god to enter their lives. To believe, as Lucretius insisted, that there were no gods, and that the world proceeded for good or ill quite indifferent to human desires, was by contrast to achieve a measure of repose through calm acceptance. Even death was not to be feared: when the atoms of one's soul and body were forever dispersed, there could be no sensation, no pain. Compared to being taunted or tortured for all eternity by frivolous, merciless gods, that was indeed a blessing.

In his philosophy, based on atomism, Lucretius found a reason to give up the struggle against blind fate, and to live instead with equanimity in the world as it was. He lived in the time of Julius Caesar, when the Roman republic was failing. Tyrants, wayward generals, and corrupt politicians would thereafter take over. Peace was to be found in withdrawing as far as possible from the vicissitudes of life. Whether Lucretius was able to live according to his own recommendation is doubtful. He suffered periods of insanity or mental disturbance, and killed himself when he was about 40 years old. In a story handed down by St. Jerome, Lucretius was so much and so often wrapped in thought that his wife grew resentful, and to restore marital relations secretly gave him a love potion. Unfortunately, the potion was stronger than necessary, drove him mad, and thus impelled him to suicide. Tennyson wrote a poem about the poet and described the reasons for his wife's unhappiness:

> Yet often when the woman heard his foot
> return from pacings in the field, and ran
> to greet him with a kiss, the master took
> small notice, or austerely, for–his mind
> half-buried in some weightier argument,
> or fancy-borne perhaps on the rise
> and long roll of the hexameter–he past
> to turn and ponder those three hundred scrolls
> left by the Teacher, whom he held divine.

This Teacher, the man by the contemplation of whose scrolls Lucretius earned his wife's displeasure, was the philosopher Epicurus, whose name survives in the notion of putting pleasure foremost among one's goals in life. A contemporary critic sniped that the ideal Epicurean way of life consisted of "eating, drinking, copulation, evacuation, and snoring," but there was more to it than that. Epicurus aimed for what might better be called contentedness, which meant freedom from pain and satiation of one's desires rather than any sort of unbridled hedonistic pleasure seeking.

To Epicurus, the greatest fear in life was the fear of death, or rather the fear of an unendurable afterlife that nevertheless had to be endured. As Lucretius reports, Epicurus employed the notion of atoms to argue that death was the final release from suffering, to be regretted, perhaps, but not feared. Lucretius differed from his teacher in one significant way: he went from atomism to atheism, but Epicurus still believed in the gods, and found the determinism of the atomic philosophy not to his taste. For that reason he introduced what seems now a rather odd idea:

> When the atoms are carried straight down through
> the void
> by their own weight, at an utterly random time
> and a random point in space they swerve a little,
> only enough to call it a tilt in motion.

Lucretius goes on to indicate that these "swerves" in the motion of atoms are what cause the atoms to cluster together or collide or otherwise interact in ways that can produce natural phenomena. The main point, however, was apparently to get around strict determinism by allowing atoms to alter their trajectories spontaneously, without any immediate cause. Perhaps this restores free will, or the ability of the gods to meddle, but it strikes the modern reader as an "unscientific" addition to the theory.

It was, indeed, Epicurus's own ill-considered addition. He did not dream up the notion of atoms, but got them from a still earlier source, in the writings of the Greek philosopher Democritus, and his teacher, Leucippus.

Of Leucippus little is known except that he flourished and taught in the years following 440 B.C. in what is now Turkey. His pupil, Democritus, lived from about that time until 371 B.C., mostly in northern Greece, and whether the beginnings of atomism should properly be credited to him or to Leucippus is impossible to say, since the latter's teaching is preserved only in the former's writings. Nevertheless, between the two of them, they put together what we can easily—perhaps too easily—see as the first intimation of a recognizably modern theory of atoms. They proposed that there exists a void, and in this void atoms move about, always in motion. Atom and void are all there is. The atoms come in a variety of distinct types and are indivisible; they band together in different ways to create the tangible and visible ingredients of the world.

To Democritus it was evident that there could be no up or down in an infinite void, and he therefore proposed that atoms move endlessly in all directions, changing course only when they ran into each other. But this implies determinism: once the atoms are off and running, their courses are fixed. There is still room for a deity at the beginning–a prime mover, an uncaused cause, or some other extraphysical influence that sets the atoms up and pushes them off in certain directions–but once that's done, determinism takes over. Does this mean there is no free will or volition? That the future is completely determined by the past? That question has haunted atomic theory, indeed physics in general, since the time of Democritus, and haunts us still today.

What distinguished Leucippus and Democritus from most of their contemporaries, and from almost all of the thinkers who followed them over the next two millennia, was that they were mainly interested in trying to understand how the world worked. Other philosophers began to focus their attention not so much on the universe as on the position of human beings in the universe, the extent to which human beings could know or understand the world around them, and how humans ought to behave. Thus arose the numerous brands of philosophy that have concerned themselves with the nature of knowledge and thought, and with the ethics and morality of human behavior. Religious philosophers took for granted that the universe has a purpose, and that humans have a purpose within it, which they may aspire to or fall away from. Leucippus and Democritus were, by contrast, scientists, aiming to understand as dispassionately as possible what is out there. Since their time, science and philosophy have become separate and frequently combative disciplines.

Atomic theory, with its implicit atheism and determinism, lost the favor of philosophical thinkers for a long period. But it crops up from time to time, for example in the writings of Isaac Newton:

> It seems probable to me that God in the beginning form'd matter in solid, massy, hard, impenetrable, movable particles, of such sizes and figures and with such other properties, and in such proportion to space, as most conduced to the end for which he form'd them.

Whether from personal belief or caution, Newton is careful to cede to God the responsibility of creating atoms in the first place. But how, if at all, is this statement an advance on anything that Democritus (through Epicurus and then Lucretius) had said two thousand years earlier? Newton lists the attributes that atoms must or might have, but then concludes, quite circularly, that the properties and behavior are such "as most conduce" to the effects they need to generate. What atoms do, in other words, is whatever they need do in order to produce the phenomena of the natural world. Neither Democritus nor Newton is able to say how, in any specific sense, atoms behave so as to generate physical effects. In the absence of any such elaboration, atomism was bound to remain an appealing but speculative picture rather than a truly scientific theory.

By contrast there were, from the earliest times, plausibly scientific criticisms of the atomic philosophy. One objection that arose in Democritus's time was later taken up with enthusiasm by Aristotle: how could atoms move constantly, without let-up, for all time? In Aristotelian mechanics, inferred from direct observation, moving objects came to a halt unless something intervened to keep them moving. You had to keep kicking a rock to keep it rolling. What, therefore, kept atoms moving?

Once Newton came along with his laws of motion, this argument lost much of its force. Newton directly contradicted Aristotle: objects keep moving, in straight lines, until something stops them. The kicked rock rumbles to a halt because the impacts it suffers sap its energy.

The other knock against atomism was that the atoms moved around in empty space, a void, and many philosophers had satisfied themselves that a void was impossible. Their reasoning, briefly, was that for anything to exist, it must have a name that referred to something rather than nothing, and since nothing by definition could not have such a name, it could not therefore exist. This argument, we would now say, is the result of a philosophical confusion between the name of a thing and the thing itself, but it took philosophers a long time to sort that one out. If indeed they have, even now.

Democritus answered these objections in essence by refusing to answer them. He simply asserted that atoms exist and that they move incessantly in the void. He didn't attempt to provide any proof of these statements, but regarded them instead as assumptions from which he and the other atomists sought to explain what they saw in the world about them.

This attitude is strikingly modern and scientific. As Democritus saw it, you have to start somewhere. You make an assumption and explore the consequences. This is exactly what scientists continue to do today, and the fact that a certain assumption leads to all kinds of highly successful predictions and explanations does not, strictly speaking, prove that the original assumption is correct. To jump abruptly to the present day, many theoretical physicists now believe that the elementary particles of the universe are creatures called superstrings–literally, lines or loops that wiggle around in multidimensional space, creating, by wiggling in different ways, electrons and quarks and photons. (More recently still, these superstrings have been subsumed into more complicated multidimensional structures called branes.) Enthusiasts for superstring theory and its variants maintain that they have hit on a fundamentally simple explanation for everything in the physical world, although working out the observable consequences of that explanation is admittedly a complicated and perhaps inconclusive business. Critics point out that whether superstring theory can be tested satisfactorily depends crucially on whether working out the details can be done, even in principle. Neither side expects that anyone will ever see a superstring in its native form.

The modern debate over superstrings is philosophically not very different from the ancient debate over atoms. To Democritus it was self-evidently a step forward to be able to explain the wildly varying and seemingly unpredictable phenomena of the natural world in terms of unchanging and eternal atoms. But even this idea had detractors. Heraclitus–famous for his observation that "you can't step into the same river twice, for fresh waters are always flowing in upon you"–believed that change, not permanency, was the essential nature of the world.

How much credit should we give Democritus and the few atomists of his era for correctly anticipating what we now know to be true? The universe is either constant in its fundamental nature, or ever-changing; matter is either continuous and infinitely divisible, or else made of a finite number of indivisible parts. There seem to be no other possibilities. On both questions, Democritus happened to choose the right side.

Then again, the early atomists were far from right about everything. They believed that the soul was composed of especially subtle atoms. Lucretius had a theory that sweet and bitter tastes arise when the tongue encounters smooth or jagged atoms. With hindsight, we tend to dismiss these errors as the products of overenthusiasm and seize on the points where the atomists got it more or less right. As Bertrand Russell put it, "By good luck, the atomists hit on a hypothesis for which, more than two thousand years later, some evidence was found, but their belief, in their day, was nonetheless destitute of any solid foundation."

What's most important about Democritus is his insistence that explanation, if it is to have lasting value, must itself rest on permanent foundations–a requirement that seems today almost a definition of what science must aim for. The Heraclitean idea that all is change and flux, on the other hand, seems to lead nowhere. Democritus, in his style of thinking, was more like a modern scientist than any other ancient philosopher. He argued that we ought to understand the universe first and worry about our place in it afterward, not adjust our view of the universe for the sake of our own peace of mind. He believed that the complexity of the world at large could, in principle, be explained by means of a simple underlying hypothesis. He believed it was not foolish to imagine the world was made of tiny components, even if those components were too tiny ever to be seen. These self-same principles, and the controversy they engendered, rose up once again almost two thousand years after Democritus, when the modern version of atomic theory began its ascent.

In that interim, atomic theory languished, never quite forgotten but not much amplified either. A taint of atheism hung over it, and natural philosophy in the post-Roman, prescientific world was powerfully religious, or at least mystical. Philosophers of the middle ages set as their most important undertaking the task of proving that God existed. The alchemists, meanwhile, tried vainly to find secret recipes that would transform, in mysterious ways, one substance into another, and in particular, base metals into gold. The towering but enigmatic Isaac Newton was in many ways both the first modern scientist and also, in Keynes's phrase, the last alchemist. When he wasn't propounding mechanical laws of motion or inventing the differential and integral calculus, Newton pored over the Bible and other ancient texts, trying out bizarre numerological schemes in the pursuit of arcane knowledge.

Nevertheless, modern science gradually emerged. The alchemists–mystics and sorcerers–changed almost imperceptibly, as pupil outgrew teacher, into chemists. Both were looking to unlock the nature of the physical world and the transformations within it, but where alchemists stumbled blindly, hoping to come across secret recipes, chemists slowly adopted a more purposeful strategy, hoping to control chemical transformations by first understanding the rules that governed them.

Atomic theory began to rise again. The rules that chemists learned imposed some restrictions that would have gravely disappointed their alchemical predecessors. Metals such as iron, copper, and gold were elemental quantities, they found, that could under no circumstances be forcibly converted one into another. Fire, on the other hand, which to alchemists had always been the supernatural agent of transformation, turned out to be just such a transformation in its own right: a chemical reaction.

Chemists grasped the idea of elements and of chemical reactions as combinations of elements changing partners according to strict rules, as in a country dance. Water, for example, was a compound of two parts of hydrogen to one of oxygen. From there it was not so big a leap to think of "atoms" of these gases combining, two of hydrogen with one of oxygen, to create an "atom" of water. (The modern distinction between atoms and molecules, which consist of several atoms bonded together, did not become fully clear until chemists had sorted out what were elements and what were compounds of those elements. In the meantime, scientists used the terms atom and molecule somewhat interchangeably.)

Still, the chemists didn't care very much (because they didn't need to) what the atoms looked like, how they behaved, how they congregated or dispersed. Whether they were tiny, hard things flying about in empty space or fat, squishy things packed closely together like oranges in a carton didn't matter much. And it wasn't at all clear whether the atoms of hydrogen and oxygen were genuine, indivisible entities, or whether the two-plus-one formula for combining them into water was simply a handy accounting method. As had been the case for Democritus and Lucretius, atoms seemed like a nice idea, at least to those disposed to think that way, but still there didn't yet seem to be anything necessary or compelling about them.

What seems surprising in retrospect, perhaps, is that it took so long for physicists to combine Newton's laws of motion, so well established as the foundation of physics in the 17th and 18th centuries, with the resurgent atomic hypothesis–to think, in other words, of atoms as little objects moving, colliding, and bouncing off each other in accordance with standard Newtonian mechanics. This is what Daniel Bernoulli first tried, in 1738, with his argument deriving pressure from a consideration of atomic motion. But even after that, in 1763, Roger Boscovich wrote an exposition called

Theoria Philosophiae Naturalis in which he offered an atomic theory that relied on essentially stationary atoms. Boscovich, a peripatetic philosopher-priest of Serbo-Croatian origins, argued that at very short range, atoms attracted each other: that was why a piece of cloth soaked up water. At somewhat longer range, however, atoms pushed each other away: that was why a gas exerted pressure.

Boscovich's account, though it has some modern elements, also illustrates why atomic theory was not taken seriously by many scientists for such a long time. Rather than imagining atoms as having certain properties, and seeking to draw conclusions about their behavior, he instead gave the atoms whatever properties he needed in order to explain the phenomena he addressed. This put into practical terms Newton's suggestion that atoms must "conduce themselves" so as to produce the behavior we see. It is easy to criticize this thinking as wholly speculative and unscientific. First you imagine that atoms exist, and then you imagine that they have whatever properties they need in order to account for the phenomena you want to explain.

These philosophical considerations aside, the other great barrier against the acceptance of atomism, especially as it applied to gases, was ignorance of the true nature of heat. At the beginning of the 19th century, opinion was divided. Some scientists thought that heat was a mechanical property of some sort, related to energy and other Newtonian concepts, but others, perhaps a majority, subscribed to the notion that heat was a kind of vaporous fluid or tenuous substance that went by the name *caloric*. This caloric was supposed to be an entity in its own right, not something composed or built from other components, and it could somehow soak into or pervade material objects, bestowing on them the property we recognize as heat. When a warm object lost heat to a colder object in contact with it, that was because caloric dribbled out of one and seeped into the other.

An argument against the caloric theory came from the Massachusetts-born scientist and inventor Benjamin Thompson, who spied for Britain in the years preceding the Revolutionary War, fled to London in 1775, returned briefly to America while the war was still going on, and after the newly independent United States had won, returned to Britain as a refugee. The appreciation shown to him there fell short of his expectations, and through political connections he obtained an appointment to the royal court of Bavaria, where he served mainly as a military adviser but succeeded in making himself indispensable in a variety of ways. He laid out the English Gardens in Munich, concocted a recipe for soup (along with specific chewing and swallowing instructions) that was meant to keep soldiers well nourished, and designed a portable coffeemaker. For these and other services he was made, in 1792, Count Rumford of the Holy Roman Empire–a name familiar to many American home renovators today in connection with the Rumford fireplace, an efficient hearth he designed in order to keep smokiness to a minimum.

Besides all this, Rumford also showed a genuine aptitude for scientific insight, and he made a number of useful observations concerning the nature of heat and energy. In his capacity as a military engineer in Bavaria he oversaw the boring out of cannons, and noticed that a dull bit would grind endlessly into a chunk of metal, achieving little except the generation of heat. He concluded that the amount of heat obtainable was essentially limitless, as long as the drill bit kept boring away. That was hard to understand if heat represented caloric being drawn out of the drilled metal; surely the original supply of caloric would run out after a while. Rumford saw instead that heat generation had something to do with the physical work of grinding the bit on the metal.

The caloric theory of heat lingered on into the first decades of the 19th century, despite observations such as Rumford's and despite the fact that no one could really say what sort of a substance caloric was supposed to be. In that respect, however, atoms–invisible particles with unknown properties–had no firmer standing. But physicists were at least familiar with gases and fluids in a general way, and if caloric was a peculiar kind of fluid, that was because heat was a peculiar kind of quantity. Atoms, on the other hand, were a complete unknown, and to explain something familiar yet enigmatic, such as heat, in terms of tiny, hard masses must have struck scientists of the early 19th century as too great a leap of imagination for them to follow.

Accustomed as we are nowadays to the idea of explaining all manner of observable or detectable phenomena in terms of remote, invisible entities–quarks and photons, electromagnetic fields, curved space, and the like–scientists of two hundred years ago were still essentially rooted in what they could see and measure directly. Heat could be detected at the fingertips; it was an undoubted physical phenomenon. The pressure of a gas could likewise be felt in the tautness of an inflated balloon or the powerful stroke of a piston in a steam engine. What did it mean to explain such direct and unarguable perceptions in terms of the undetectable actions of invisible objects? Certainly, one could imagine tiny atoms bashing against a piston head, and mustering enough collective force to push it out, but what was the advantage in imagining such a thing? As an explanation, this seemed to be going the wrong way, portraying something immediate and tangible in terms of "atoms" that were forever concealed from the human eye. Although Bernoulli, then Herapath, then Waterston had made more precise and specific the atomic picture devised by Democritus and recounted by Lucretius, they had not yet improved greatly on the nature of the argument: a critic could still argue, with considerable reason, that this was a nice picture, but hardly a scientific theory. It might explain one thing in terms of another, but not yet in a sufficiently broad way to make physics overall any simpler.

As late as 1845, when John Waterston submitted his ill-fated manuscript to the Royal Society, the explanation of heat as atoms in motion–it was known as the kinetic theory of heat–was to some extent a hypothesis in search of a problem. If you were inclined to believe in atoms in the first place, kinetic theory seemed like a pleasing extension of a broad and fundamental picture of nature. But if you were disinclined to atomism, the kinetic explanation didn't seem to say anything you didn't already know.

And yet, in just 12 years, kinetic theory went from outlandish idea to respectable proposal. It was not so much that the theory was suddenly improved, or found able to explain wholly new matters, but rather that a handful of influential people began to take it seriously. In 1857, the German physicist Rudolf Clausius, already well known for his work on the relationship of heat and mechanical energy, published an influential work titled *The Kind of Motion We Call Heat.* Clausius was 35 years old at the time, and his reputation was solid but not yet remarkable. He had been working for some years in opposition to the caloric theory of heat, trying to prove instead that heat was, as Rumford's observation had indicated, intimately linked to mechanical work and energy. He said, in 1857, what Bernoulli and Herapath had hinted at, and Waterston had propounded in some detail. If a volume of gas consists of tiny atoms in relentless motion, then both the pressure it exerts and its temperature are related in a simple way to the square of the average velocity of the atoms. Temperature, in fact, is nothing other than the average kinetic energy of these presumed atoms.

It is hard, in retrospect, to see why Clausius's work was taken so much more seriously in 1857 than Waterston's had been a dozen years earlier, except that Clausius was an established professor of physics at the polytechnic institute in Zurich, while Waterston was an instructor at a naval college in Bombay. In both the German-speaking and English-speaking worlds, several notable physicists had become convinced that heat was ultimately mechanical in nature, and in fact controversies erupted from time to time over whether Clausius or his British rivals had propounded certain ideas first. It was true that between 1845 and 1857 quantitative laws had been enunciated (by Clausius, among others), making the connection between heat and energy more precise, and it may be that the passage of 12 years was just enough to carry the kinetic explanation of heat across the threshold of credibility, from speculative suggestion to scientific proposal.

Clausius, at any rate, is the man whose work brought the atomic theory of heat into the scientific world. With his imprimatur, kinetic theory was taken more seriously and attracted new adherents. Younger scientists saw it as an enticing idea and worked to improve it. In 1860, James Clerk Maxwell published in England an elaboration of Clausius's theory, taking into account not simply the average speed of the atoms, but their distribution of speeds as well–that is, how many, at any given time, are moving at speeds greater or smaller than the average. He derived, on somewhat abstract and not entirely persuasive grounds, a mathematical form for this distribution, in effect a graph of the typical speeds of atoms in a volume of gas at any given temperature.

Maxwell was at that time only 28 years old, but signs of his brilliance were already apparent. The next step came from a still younger man, and one whose name was then known to hardly anyone. In 1868, the 24-year-old Ludwig Boltzmann, newly graduated from the University of Vienna, published a more convincing physical explanation for the formula Maxwell had derived. By analyzing what would happen to a volume of gas rising in Earth's gravitational field–a case for which the change in pressure with altitude was already well understood–Boltzmann showed that Maxwell's formula correctly predicted how the number of atoms or molecules with a particular energy would correspondingly change. This was a remarkable stroke of insight for a young man who had only recently finished his studies.

Boltzmann's argument gave a direct and easily grasped physical justification for Maxwell's formula, and showed moreover that there was real physics in the new kinetic theory. Maxwell himself was impressed, and wrote to one of the senior Viennese physicists expressing his admiration for Boltzmann's work. The formula that these two young physicists proposed became known as the Maxwell-Boltzmann distribution for the velocities of atoms in a gas, and it remains the cornerstone of the atomic depiction of gases. Clausius, the somewhat older man, set kinetic theory on the path to respectability, and Maxwell made fundamental contributions when he was not occupied with other theoretical endeavors. But it was Boltzmann who made the full development of kinetic theory his life's cause and who took its missteps as much as its successes on his shoulders. During the second half of the 19th century, the peaks and depths of Boltzmann's difficult life mirrored exactly the stumbling ascendancy and frequent reversals of kinetic theory itself.

In due course, Daniel Bernoulli's forgotten century-old proposal was rediscovered, and his insight was recognized by Boltzmann and everyone else. Herapath, after his early forays into kinetic theory, had published some letters in the London *Times* attacking Davy, but he then focused his energies in other directions. He went briefly and unsuccessfully into teaching, became interested in the blossoming railway industry, and in the end became a well-known writer and commentator on that business. He remained an amateur scientist and published a few small works in *Railway Magazine and Annals of Science,* of which he was conveniently the editor. Maxwell, much later, acknowledged that Herapath had had the right general idea but observed that his calculations of atomic collisions were incorrect.

Waterston's story has no happy ending. His rejected paper of 1845 contained the essence of what Clausius shortly afterward became famous for, and contained hints too of many other ideas that were not fully developed until some time later. A brief abstract of his paper appeared in the Royal Society's *Proceedings* in 1846, and another short notice was published in 1851, but these accounts were so brief, and gave so little indication of the conclusions Waterston had reached, that they went unremarked. His original manuscript was not returned to him but languished for decades in the files of the Royal Society. After he had returned from India to Scotland he published scientific papers here and there without ever becoming much known to his contemporaries. In 1878, he resigned from the Royal Astronomical Society after two of his papers were rejected for publication. Thereafter, according to a nephew, any mention in Waterston's presence of the scientific societies "brought out considerable abuse without any definite reasons assigned." He died in 1883, at the age of 72.

Waterston's achievement finally came to light in 1891, when the English physicist Lord Rayleigh, then secretary of the Royal Society, discovered the lost manuscript in the course of tracking down some old citations. By that time the kinetic theory amounted to a sophisticated and well-known body of knowledge, and Rayleigh immediately perceived the true merit of Waterston's ideas. He arranged for its belated publication as the first item in the first issue of the *Philosophical Transactions* for 1892, along with a brief commentary on its tortured history.

Acknowledging that Waterston had submitted his work at a time when scientists thought very differently than they were accustomed to doing just a few decades later, Rayleigh admitted nevertheless he was surprised that the Royal Society's expert reviewers were so dismissive of the paper. "The omission to publish it at the time was a misfortune, which probably retarded the subject by ten or fifteen years," he wrote. Rayleigh suggested that Waterston might have done better to mention that he was working to elaborate ideas previously suggested by Daniel Bernoulli, whose reputation was unarguable; that might have made a reviewer hesitate. But Bernoulli's work had itself been forgotten, and it is the strength of Waterston's claim to unjust treatment that he indeed came up with his reasoning entirely by himself. On that score Rayleigh had another observation: "Perhaps . . . a young author who believes himself capable of great things would usually do well to secure the favourable recognition of the scientific world by work whose scope is limited, and whose value is easily judged, before embarking on greater flights." The reliable route to scientific fame, in other words, requires brilliance judiciously combined with careerism. Just as well, perhaps, that the already bitter Waterston didn't live to see this endorsement of his unrewarded endeavors.

CHAPTER 2

Invisible World

The Kind of Motion We Call Heat

DEMOCRITUS HAD NO WAY OF KNOWING how big an atom might be. Nor, almost two thousand years later, did Bernoulli, Waterston, or Clausius. Atoms had to be invisible, but by the standards of the mid-19th century that meant only that they must be smaller than, say, a grain of pollen. But how much smaller? A hundred times? A million?

When Ludwig Boltzmann entered the University of Vienna in 1863 as an undergraduate, he found himself quite by chance at an institution where the new atomic thinking was being enthusiastically embraced. An exceptionally bright young man from a provincial school, he had as yet shown no leaning toward any particular aspect of physics, but his teachers and mentors in Vienna were among the most forward-thinking in continental Europe. They knew of the latest developments in kinetic theory and electromagnetism, and their early eagerness for these theoretical innovations set the young Boltzmann on a path of intellectual adventurousness that was to sustain him throughout his life.

Just two years after his arrival in Vienna, an older scientist there used some of the ideas of kinetic theory to produce the first credible estimate of the size of atoms. Though he was in his forties at the time, Josef Loschmidt was, like Boltzmann, new to research; he had only recently returned to his youthful interest in science after an erratic career in business and teaching. Born to a peasant family near what was then Carlsbad in Bohemia (now Karlovy Vary in the Czech Republic), Loschmidt had been a bright youngster. A local priest encouraged him in his schoolwork, and he showed such loathing and incompetence for working in the fields that his parents eventually gave up on him as being good for nothing but studying. An early interest in philosophy was steered toward the sciences when he came under the influence of the physicist Franz Exner at the University of Prague, and when Exner moved to a new position in Vienna, Loschmidt followed.

But he was unable to find a teaching position in Vienna. Loschmidt instead harnessed his scientific knowledge, especially of chemistry, to a series of business ventures, which repeatedly and reliably failed and took him into bankruptcy in 1854, at the age of 33. He decided that perhaps his true talents lay in the realm of philosophy and reason after all. Boltzmann later described Loschmidt, despite the time he spent in the industrial world, as "the prototype of the impractical academic." He obtained a certificate to teach sciences in Vienna and began to involve himself actively once more in science. Not until he was 40 years old did Loschmidt publish his first original scientific research.

Boltzmann, a young man newly arrived in the capital city, developed a close relationship with the amiable Loschmidt. They went to the opera and symphony performances together. Vienna was preeminently a city of music, and Boltzmann was a talented pianist. His first experience of the Vienna Philharmonic was a performance of Beethoven's Third Symphony, the *Eroica,* which he attended with Loschmidt. Boltzmann later recalled that he tried to impress the older man by being "especially clever. Instead of the scherzo, I said, a movement representing the hero's fate with solemn seriousness would have suited me better."

Loschmidt was quick to admonish his young companion. "So,

you would have done a better job than Beethoven! Have you ever been to the funeral of a great man, whom you adored? Did you see him ascend to heaven? No! But you had to go back to daily business, which then appeared twice as stale to you, so you had trouble suppressing a loud and scornful laugh–that is the scherzo!" Boltzmann recalled these words four decades after the event, and one wonders how much they represent Loschmidt, and how much Boltzmann himself. Throughout his life his artistic tastes inclined to the overtly dramatic. He loved Beethoven's symphonies and played them on the piano in Liszt's transcriptions. Later he became a devout Wagnerian, and in the theater he was a devotee of the German playwright Friedrich Schiller, who at least in his earlier years was a conscientious pupil of the Sturm und Drang school, emphasizing raw feeling and unfettered emotion above irony or introspection. In Loschmidt the young Boltzmann found a fellow enthusiast for German romanticism in its ripest manifestations.

Loschmidt acted as a sort of father figure as well as a companion to the young Boltzmann, whose own father was no longer alive. When they were not at the concert hall or the theater, they would go to the beer halls in the evening to talk about science, both eager to explore the new world of research that was opening before them. Loschmidt's first ventures mainly concerned the structures of chemical compounds. It was his innovation to use double and triple lines to represent, in chemical diagrams, double and triple bonds between atoms, a pictorial system in standard use today.

With his interest in chemical structures, it was natural for Loschmidt to latch onto the new atomic thinking, which, in the late 1850s and early 1860s, was beginning to gain some currency. Despite the work of Clausius and Maxwell in elucidating the velocities of atoms and their relationship to the physical properties of gases, little could be said directly about the atoms themselves. There was no way to tell, for example, how big they were: a lot of tiny atoms would exert the same overall effects as a smaller number of bigger atoms. It was this kind of uncertainty and ignorance that gave critics ample reason to dismiss the atomic hypothesis as groundless speculation. If the same results ensued whether the atoms were large or small, the whole idea seemed far too arbitrary to be taken seriously as science.

At first glance, the advent of kinetic models of gases seemed to make this ignorance even greater. One supposition had been that atoms were bloated, spongy entities, packed together like a box of tennis balls or a stack of oranges. A gas was compressible because the atoms themselves were soft and exerted pressure because of repulsive forces between them. In such a model there was a direct relationship between the size and number of atoms and the volume of space they would collectively occupy. But in the new kinetic theory, atoms were zooming about in empty space, not nestling up against each other. They could even be pictured in terms of that old mathematical standby, the idealized point, possessing mass and speed but having no meaningful dimensions. In the new atomic models, it appeared, size didn't matter.

But as physicists met and overcame objections to the earliest incarnations of kinetic theory, they found they had to be more specific about what atoms were. In *The Kind of Motion We Call Heat,* Clausius had shown how to relate the temperature and pressure of a volume of gas to the motion of the atoms, and was able to deduce their average speed. It turned out to be some hundreds of meters per second, perhaps even a thousand meters per second. That calculation drew a quick response from the Dutch meteorologist Christopher Buys Ballot. He observed that if he were seated at one end of a long dining room and a butler brought in dinner at the other, it would be some moments before he could smell what he was about to eat. If atoms were really flying through the air at hundreds of meters per second, shouldn't the fragrant vapors of a hot dinner race through the room in an imperceptibly small instant, so that he would smell the dinner as soon as he saw it?

In figuring out the answer to that puzzle, which he soon did, Clausius added a fundamentally important innovation to gas theory. Atoms moved very fast, it was true, but they also banged into each other a good deal. An atom would make its way from one end of a dining room to the other not by cruising unimpeded along a straight path, but by battling through all the other atoms like an opera-goer trying to push through the crush to get out of the theater at the end of the show. Each atom would travel a little way, collide, go off in some other direction, then collide once more. Atoms might move quickly between collisions, but their overall progress was dictated by how many zigs and zags they took to get from one place to another. What mattered was the average distance between collisions. This turned out to be an all-important quantity in kinetic models of gases, and Clausius gave it the name *mean free path.*

Implicit here was a further observation: if atoms were to collide, they must have some physical size. Mathematically idealized points would always slip past each other no matter how close they approached. So Clausius now imagined atoms as tiny hard spheres–the "billiard ball" atom that turns up in so many handy analogies. The bigger the sphere, the more likely atoms were to run into each other, and the shorter, therefore, their mean free path.

In 1858, when Clausius introduced this new element to the theory, he could only say that there ought to be some finite mean free path for any particular gas, but he didn't have any way of putting a specific number to it. The size and number of atoms remained unknown. Clausius's picture of gases as aggregations of moving atoms was becoming more physically detailed, but in its quantitative details it remained sadly vague.

The following year, the Scotsman James Clerk Maxwell provided an elaboration of Clausius's ideas that included an unexpected vindication. Maxwell was a sharper mathematician than anyone who had tackled the subject thus far, and he was not awed by its complexity. In their calculations, Clausius (and Waterston, for that matter) had imagined all the atoms in a gas moving at the same speed. They knew this wasn't true, that in fact atoms would move with a range of speeds, but they didn't have the mathematical sophistication to tackle the full problem. Maxwell, however, knew just how to proceed. He defined a mathematical function called the distribution of velocities, which kept track of how many atoms were moving at any particular speed relative to the average, and by dealing in terms of this distribution rather than a single assumed average, he was able to give his calculations a precision that those of Clausius lacked.

Maxwell then set himself to figuring out from his model another standard physical property of a gas, its viscosity. This quantity marks the ease with which a gas or liquid flows or, conversely, the resistance it puts up to the movement of a solid object through it. Water isn't very viscous; molasses is.

To his considerable surprise, Maxwell found that for any particular gas, the viscosity predicted by kinetic theory ought to be independent of its density. This seems opposite to what one would expect: an object trying to move through a denser gas has to push more atoms out of the way, and so it would seem obvious that there should be more resistance to its passage. The reason behind Maxwell's unexpected result is a subtle one. Viscosity arises because the constant bombardment by atoms saps energy from an object moving among them. What matters, Maxwell found, is the number of atoms within one mean free path of the object's surface: as the density of a gas falls, the mean free path increases, and the moving object interacts with atoms coming at it from farther away. The decreasing density and the increasing mean free path compensate for each other, so that the number of atoms within one mean free path of the object remains the same–as does the viscosity.

So taken aback by this result was Maxwell that at first he thought he had found a refutation of Clausius's new-fangled kinetic theory. A Cambridge colleague he consulted told him that what little evidence there was on the matter seemed to indicate, as common sense would have it, that less dense gases were indeed less viscous. In his first publication on the subject, Maxwell showed how to calculate viscosity and noted that the predictions seemed to go against what was known of the actual behavior of gases. But he then decided to do some careful viscosity measurements himself. To his further consternation, he now found that measured viscosity indeed seemed to stay the same over a wide range of gas density. The earlier experiments were, as his colleague had told him, somewhat sketchy, and Maxwell reckoned his own work was more reliable. Now he began to think he had proved Clausius correct, the more so, since kinetic theory implied a surprising consequence, which then turned out to be true.

By 1866, when he published a long analysis including these new results, Maxwell had become convinced of the value and importance of the kinetic theory of gases. He was thereafter one of its great supporters and innovators, and spearheaded its acceptance and development in the English-speaking world.

In the meantime, Loschmidt seized on another use for the viscosity calculation, which connected the mean free path of the atoms or molecules in a gas to a measurable property of the gas as a whole. As always, both the size of the molecules and the number of them in a given volume entered into the viscosity, but they did so, Loschmidt noted, in a novel way. If he could compare the viscosity calculation to some other formula involving the same two quantities he would have, as the mathematicians put it, two equations for two unknowns, and the problem would be solved.

Loschmidt reasoned that in a liquid, the atoms or molecules would be more or less squeezed up against each other, so the volume of a liquid would be straightforwardly the volume of an individual molecule multiplied by the number of them. Moreover, standard experimental data provided a conversion factor for the volume of gas that a given volume of liquid would create when evaporated. In this way, Loschmidt arrived at an estimate of the size of a typical molecule in air. The diameter he came up with was a little less than one millionth of a millimeter–by modern standards a pretty fair answer.

To enthusiasts for atoms and the kinetic theory of gases, a quantitative estimate of the dimensions of atoms was yet another piece of evidence that they were on the right track. Showing that the invisible objects must have a particular size made them, perhaps, a little more real. To critics, on the other hand, Loschmidt's analysis still didn't prove anything. The fact that some mathematical expressions and experimental data could be knitted together in such a way as to yield a number corresponding to the diameter of a purely hypothetical object was in no way evidence that the hypothetical entities were real. Loschmidt's algebra at least proved that the mathematics of atomic and kinetic theory was consistent within itself, a minimum expectation for any theory. But the argument against kinetic theory was that in the absence of tangible evidence that atoms existed, it was mere mathematics, empty theorizing. Loschmidt had shown that if atoms existed, they must have a certain size–but that first "if" had not been overcome.

There was, for the time being, no good rejoinder to this skepticism. Nevertheless, the atomists perceived, or perhaps simply believed, that accounts based on atomic theory could ultimately be more complete, more encompassing, more intellectually satisfying than a set of empirical laws deduced from experiments and otherwise left unexplained. But such notions, only dimly apparent, were rather new to physics. Traditionally, the subject had concerned itself essentially with searching out quantitative relationships between measurable phenomena: the laws relating the temperature and pressure of gases were the perfect example. To go beyond this, to explain observable facts in terms of unobservable but allegedly "real" entities such as atoms, was to go beyond what many physicists regarded as the limits of their discipline. What was happening, in the second half of the 19th century, was the birth of the subject we now call theoretical physics as a discipline in its own right, intimately linked to experimental physics (or so one would hope) but at the same time distinct. The notion of a theory of physics as a sort of free-standing intellectual structure, related to facts and observation but at the same time employing theoretical constructs that found their true definition only within the confines of that theory–this was a puzzling innovation. How was the physicist to decide whether the theoretical constructions were real? Was it enough that they were simply useful or illuminating in some way?

As well as having a hand in kinetic theory, Maxwell was more or less singlehandedly responsible for the other great theoretical innovation of his time. In 1864, he published his theory of electromagnetism, which elegantly and comprehensively showed how all the observations and inferences concerning electricity and magnetism could fit under a single roof. Maxwell's theory introduced a new idea, the electromagnetic field. Like atoms and molecules, this field could not be observed in a direct way, but as a theoretical structure it underpinned the numerous ways that electric and magnetic phenomena were known to be related. The theory held that light was an oscillation of the electromagnetic field, and it predicted that other wave motions should also arise. It was not until 1888 that Heinrich Hertz, in Germany, generated and received radio waves in his laboratory, verifying an essential aspect of Maxwell's theory. In the meantime, the theory had supporters and detractors. As with the kinetic theory of gases, critics stumbled over the idea of theoretical creations that seemed intrinsically undetectable. Over atoms and electromagnetic fields, the same question arose: real or imaginary?

Loschmidt published his estimate of the size of air molecules in 1865. It was an encouraging sign, but not proof that atoms existed. Still, and for some years yet, believers in atomic theory had to rely on little more than an instinct that they were on the right path, or perhaps simply an interesting path, worth following even if it eventually ran into a dead end.

As an undergraduate at the University of Vienna, Boltzmann could hardly have failed to be excited by the new physics that was materializing around him. It was a propitious time and place. The Institute of Physics in Vienna had been in existence only since 1850, but it was already undergoing a rebirth. The institute's founder and first director had been Christian Doppler (he of the Doppler effect, which makes train whistles sound at a higher pitch when the train is approaching). But Doppler had died in 1854, only 51 years old, and the institute had then been run by Andreas von Ettingshausen, a physicist of no more than average competence, who was actually a few years older than the man he succeeded. In 1862, von Ettingshausen had become ill, and though he remained formally the director, a younger man was sought to take charge of running the institute. Vying for the position were the 24-year-old Ernst Mach and another up-and-coming physicist, Josef Stefan, who was three years older.

Both Mach and Stefan were young men with plenty of promise but, as yet, no memorable achievement. That two such men were being considered for an appointment of some significance testifies both to their qualities and to the modest reputation of the Institute of Physics at the time. Mach had been von Ettingshausen's student, obtaining his doctorate in 1860, but Stefan got on better with several of the other faculty members in Vienna. Stefan was chosen to take over the day-to-day running of the institute, and when the ailing von Ettingshausen finally retired in 1866, Stefan succeeded to the director's position and remained there until his death a quarter of a century later. Mach, meanwhile, who was struggling to make ends meet by lecturing, eventually found himself a position as a professor of mathematics at the University of Graz, Austria's second largest city, some 100 miles southwest of Vienna.

Stefan was a forward-thinking man. He embraced the new discoveries in atomism, and it was because of his influence that Loschmidt came to have a secure academic position. After his business failures and move into teaching, Loschmidt had become an unofficial hanger-on at the university, using the library and laboratories through Stefan's favor as and when he could find the time. When Loschmidt began to publish well-regarded work, Stefan was able to move him into a junior professorial position in 1868. A few years later he gained an honorary Ph.D., and he remained at the University of Vienna for the rest of his life.

Stefan also saw in Boltzmann a young man of great intellect and potential. An early enthusiast of Maxwell's electromagnetic theory, Stefan took care to see that his young protégé was aware of the significance of this great advance. He gave Boltzmann copies of Maxwell's papers and an English grammar, and told him he would do well to study these ideas. Boltzmann recalled later that he knew no English, but armed with Stefan's grammar and a dictionary his father had given him, he succeeded in familiarizing himself with Maxwell's new theory. And he credited Stefan with being one of just a few scientists on the continent who perceived from the start the enormous revolution in physics that the new electromagnetic theory embodied.

Equally important was the atmosphere that Stefan engendered. When he took over day-to-day administration of the Institute of Physics, Stefan was just 28 years old–hardly older than the incoming students. Born near Klagenfurt in the southernmost part of modern Austria, he was the son of illiterate Slovenian peasants but had risen quickly through grammar school in Klagenfurt and on to the University of Vienna, becoming a physics lecturer there when he was just 23 years old.

Stefan was, by all accounts, an amiable and unpretentious man. He treated his students–especially the brighter of them, no doubt–as colleagues. Boltzmann later remembered his "olympian cheerfulness" and his informal manner, so that conversations between students and teachers at the institute were as among friends. This bonhomie, perhaps arising from a remembrance of Stefan's own humble origins, set a pattern that Boltzmann was dismayed to discover was not the norm at all universities. "It never occurred to me that it was not proper for me, a student, to adopt this tone," he remarked later, but his subsequent experiences, especially farther north in the German-speaking world, made him see otherwise.

The institute was at that time in cramped accommodations at 19 Erdbergstrasse in Vienna, and in later years when he was in charge of his own physics departments and institutes, Boltzmann strove but never quite managed to recreate that fondly recalled ideal in places he would call "Little Erdberg." He would lament how his students, housed in capacious laboratories and blessed with modern equipment, would come to him asking what they should do; in Erdberg, he said, he and his fellow students never lacked for ideas, and only encountered difficulty trying to rustle up the equipment they needed. For Boltzmann, Erdbergstrasse was the occasion of joyful discovery, as the world of physics opened up before him, and no matter what success he achieved in later years, he was never able to fully recapture that blithe sense of wonder.

BOLTZMANN'S ENROLLMENT as an undergraduate in Vienna in October 1863 marked his return after many years' absence to the city of his birth. His father, Ludwig Georg, was a tax official, a middling bureaucrat in the complex and far-reaching bureaucracy of Imperial Vienna. Soon after Ludwig's birth in 1844, his father was transferred to the provincial Austrian city of Wels, and later to Linz. In Linz, roughly 100 miles east of Vienna, Ludwig received his first formal schooling, having been educated privately until the age of 10. The Boltzmanns were hardly wealthy, but Ludwig's mother, Katharina, came from a moderately affluent mercantile family in Salzburg, and the salary of a tax officer–Boltzmann senior was by this time a "regional finance commissar"–was evidently sufficient to support some of the trappings of middle-class living. For a time young Ludwig took piano lessons from the 30-year-old Anton Bruckner, who was then the cathedral organist in Linz and had yet to establish himself as a composer. The lessons ended when Bruckner left a wet raincoat on a bed and was scolded by Boltzmann's mother, but the boy was a gifted pianist and continued to play with impressive skill throughout his life despite, as one colleague later recalled, his "stubby fingers and pudgy hands."

There was no history of intellectual achievement in the Boltzmann family. Ludwig's grandfather, originally from Berlin, had been a clock maker. Artisans and craftsmen were much in demand in Vienna in that era. The industrial revolution was slower to catch on in Austria than it had been in many other places, and Vienna had a preeminent business of its own: running the far-flung and rambling enterprise known as the Austro-Hungarian Empire. Vienna was home to the monarchy, its attendant aristocracy, and a sprawling bureaucracy. The gentlefolk needed furniture and porcelain and expensive knick-knacks to fill their apartments. Indeed, smart living accommodations in Vienna in the late 18th century onward were generally not capacious, since the city proper was still constricted within its encircling military fortifications. The well-off and well-born Viennese stuffed their cramped apartments with as much opulence and display as they could lay their hands on. A skilled clock maker could therefore make a good living, and by that means give his son an education and steer him into the Habsburg bureaucracy–a good thing, as it happened, since during the first half of the 19th century small factories were beginning to displace the artisans, but there was ever a demand for civil servants.

Ludwig Eduard was born outside the city proper, in the Landstrasse suburb, then home to a mixed population of Austrians, Serbs, and Czechs, speaking a variety of languages, many of them engaged in small businesses, and many of them by that time beginning to struggle financially as industrial enterprises took away the living of craftsmen and large retail stores undermined small shopkeepers. Here, as across Europe in general at the time, social discontent and the twinges of future revolution were in the air.

The Boltzmanns, however, decamped to smaller provincial towns before anything untoward came to pass. In Linz, young Ludwig proved to be an unusually bright child, and with the exception of one year when he was held back by a debilitating cold or flu, he was always first in his class. He took piano lessons, collected beetles and butterflies, and built himself a little herbarium; later on, he encouraged in his own children an interest in music and nature.

But if there were, in Boltzmann's childhood and early education, any signs of incipient genius, they remain concealed. Many scientists in their later years recall a moment of youthful illumination or epiphany, when some striking discovery or observation sparked a lifelong fascination with science, and set them–so it seems in retrospect–on their life's journey. Not so with Boltzmann, apparently. He himself never recalled any such incident, any episode that made it clear to him that he was destined for a life of intellectual discovery, whether in physics or in any other subject.

Nor has any school companion recorded memories of the young Boltzmann, of his appearance or character or intellect. He was evidently diligent (he attributed his subsequent poor eyesight to the hours he spent studying by the light of a candle), but he seems as a young scholar to have been good at his studies without showing any particular sign of enthusiasm for one subject or another.

The tiniest sliver of a clue to Boltzmann's intellectual awakening comes from a lecture he delivered toward the end of his life, when he recalled a conversation he had had with his younger brother, Albert, when both were still in their early teens. Ludwig wanted to believe that all knowledge could be presented systematically, as long as every new idea or concept was clearly defined when it was introduced. His brother argued otherwise and eventually persuaded him that this could not be so. He asked Ludwig to imagine trying to read Hume in English, a language he didn't understand; even with a complete dictionary specifying the meaning of each word, he still wouldn't be able to fully grasp Hume's meaning. There is more to knowledge, Albert argued to Ludwig, than a mere list of definitions.

It's notable that the two boys were discussing Hume and the nature of knowledge at an early age. More notable is that Ludwig evidently wanted to insist on the orderliness of knowledge, which seems more the attitude of a thorough but unimaginative student than of someone who was to revolutionize physics. Boltzmann, it appears, was a man whose individual genius was to awaken later.

His early life was not all plain sailing, nevertheless. Boltzmann's father died in 1859, probably of tuberculosis, when Ludwig was 15 years old. Albert, his brother, died the following year at the age of 14, probably also of tuberculosis. Some years later his younger sister, Hedwig, remarked to Ludwig's future wife that after the death of his father he was "always serious." Even this small remark is hard to interpret, since there is no recorded evidence of a lighthearted, jolly young Boltzmann beforehand. And if Boltzmann had thoughts about his deceased father and brother beyond that, he kept them to himself. In the middle of the 19th century, many children lost a parent while still young, and many siblings never made it to adulthood. Boltzmann's losses, whatever their effect on him, were hardly unusual for the times. The loss of a working father's income straitened the family somewhat, but there was a pension, and the family coped without undue hardship. Boltzmann's mother now devoted her sole attention and the family's resources to the education of her precocious son.

When Ludwig returned to Vienna to study, mother and sister came too, and the three lived there as a family. There had been a revolution in their absence. In February 1848, workers and students had taken to the streets in Paris, and within weeks the uprisings spread across continental Europe. Resentment against autocratic rule was the mainspring of rebellion everywhere, and liberalization and constitutional reform of varying degrees came about. But across German-speaking Europe and the Austro-Hungarian Empire, nationalism was never far from the surface. An abortive attempt to unite the numerous German cities, states, and duchies into a confederation fizzled out, in part because Austria was not at all willing to abandon the non-German parts of its dominion and throw its lot in with the rest of the German world.

To mollify the crowds in Budapest and Prague, however, the Habsburg government was obliged to make small concessions of political power to the Hungarians and Czechs, setting a pattern that was to recur during the coming decades. The Austro-Hungarian Empire was not any sort of natural geographical or nationalistic realm, but a haphazard collection of Germans, Hungarians, Czechs, Poles, Serbs, and many more, whose only unifying principle was the ruling family itself, the Habsburgs. Before the revolution, Austria-Hungary was a thoroughly centralized entity, administered strictly from the royal court in Vienna. Its workings were so closely guarded that the English politician Disraeli referred to it as "the China of Europe." There was censorship and a pervasive secret police force; political dissent encompassed anything that was deemed injurious to the interests of the Habsburgs. Its citizens were held together not by any collective sense of national identity or common destiny, but by allegiance to the ruling family. It was a Habsburg emperor, Franz I, who on being told that a certain public official was a great patriot came up with the rejoinder, "Ah, but is he a patriot for me?"

Despite political repression and autocracy Austria-Hungary was not, for most of its people, a harsh place to live. Faced with the problem of running an empire composed of numerous peoples speaking many languages, the essential element of the Habsburg political philosophy, insofar as it had one, was to have everyone get along as well as they could. The central preoccupation of successive administrations in Vienna was the endless power struggle with the other states and nations of Europe. Domestic policy was largely a matter of maintaining a reasonable calm. The music and opera for which Vienna became famous were seen quite overtly as useful diversions; as long as people were filling the theaters, they wouldn't be in the streets. Prerevolutionary Austria was at the height of the so-called Biedermeier period, named after a complacent fictional character who embodied the times; prosperous and happy, he enjoyed his pleasant life and had not a political thought in his head.

Biedermeier Austria was in a number of ways a surprisingly egalitarian place, not for ideological reasons but because it was good for the smooth order of the empire that talented people should find fulfilling occupations and not come to feel stymied or held back. Men such as Josef Loschmidt and Josef Stefan, bright people from impoverished backgrounds, were able to find and create opportunity. That they were able to obtain an excellent scientific education was also the result of a political decision. Encouraging young people to learn science was seen, in its way, as a counterpart to encouraging an interest in the opera; as long as they were learning the hard and objective facts of science, they would have no time or energy for nonconformist philosophy or political free-thinking. As one political adviser put it, "Anyone could philosophize as the spirit moved him, but positive sciences had to be learned."

The sciences, therefore, were encouraged for a largely negative reason, in that the Habsburg bureaucracy, ever fearful of dissent and dissatisfaction, reckoned that learning science would both subdue intellectual independence and divert intellectual energies in a harmless direction. This stands in odd contrast, for example, to the role of science in the former Soviet Union, which was the breeding ground for dissenters such as Sakharov and Scharansky. The Soviets did their best to wrap science in a Marxist straitjacket and must have been constantly dismayed when scientists, despite their ideologically directed educations, started to have ideas of their own. The independence of science, its insistence on understanding the world as it actually is, strikes us today as an antidote to the kind of enforced ideology that Soviet rulers wished to make the world conform to.

But the Habsburgs, unlike Lenin and his successors, had no guiding ideology save for a general inclination that the world should be ruled by their sort of people. Science, if it taught that the natural world obeyed a certain preexisting and unarguable order, seemed to be saying the same thing. The revolutionary activity of 1848, on the other hand, posed a great threat to order, and so fearful was the Viennese aristocracy of dissent, nationalism, and anarchy that even small concessions to democracy seemed inadmissible. Twice during 1848 the court fled Vienna, returning for good only after brief but harsh military action had finally put down the revolt in November. In the next couple of years, revised constitutions with modest expansions of parliamentary powers were drawn up, discarded, enacted, and for the most part ignored.

In the end, the monarchy held onto its power, with the help of a newly installed emperor. When students and workers had filled the streets, the emperor was Ferdinand, the son of Franz, who had died in 1835. Ferdinand was generally regarded as a simpleton, but as long as times were peaceful, the empire ran smoothly enough under his disengaged rule. The revolution tainted his authority, however, and when order was finally restored to Vienna the monarchy was in the hands of Ferdinand's nephew, Franz-Josef, who had just turned 18. He proved to be a durable monarch. The biggest changes engendered by the revolts of 1848 were not in parliamentary reforms, which Franz-Josef contrived to go along with when it suited him and ignore when it did not, but in the dispersion of political powers to other parts of the empire. Demands for a degree of political representation or self-determination by the Hungarians, Czechs, and others were the bane of Franz-Josef's monarchy. To keep the peace, various ethnic groups had to be ceded limited powers, but every time more power was given away, the empire itself gave up a little more of its raison d'être.

The most evident signs of change in Vienna after the revolution were physical. The city had retained a wide swath of military fortifications and earthworks around its old center, with newer suburbs springing up beyond. The increasing scarcity of living quarters, coupled with the realization that the biggest threat to Vienna's security was no longer necessarily from without, led Franz-Josef to have the old ramparts leveled and a sweeping new boulevard, the Ringstrasse, take its place. This great avenue contained old Vienna on three sides, with the Danube canal forming the fourth. From the late 1850s to the early 1870s, numerous grand apartment buildings and a selection of great public edifices sprang up along the Ringstrasse, including the royal theater and opera, museums, a council hall, and new university buildings.

The Boltzmanns returned to Vienna in 1863, while this vast reconstruction was in progress. Despite the upheavals of 1848, the city remained a comfortable place for a middle-class family, especially a German-speaking one with no political interests. Boltzmann's education in science progressed easily. He became a doctor of philosophy in 1866, at the age of 22, and was immediately taken on by Stefan as his assistant. And just as quickly he began to publish original scientific research. His very first paper, published in 1865, was a short note "On the motion of electricity in curved surfaces." The problem he solved here had arisen in the course of Stefan's lectures, and Boltzmann was impelled to publish his solution after coming across a recent textbook that gave the wrong answer.

His attention very quickly and fruitfully turned toward kinetic theory. He published a couple of short articles on aspects of atomic motion in gases, but it was his first substantial article of work, a 40-page analysis published in the *Proceedings of the Viennese Academy of Sciences* in 1868, that demonstrated his ability to tackle and resolve problems at the forefront of research. This was Boltzmann's landmark demonstration that Maxwell's formula for the distribution of speeds and energies among a collection of atoms was not only mathematically sensible but also physically reasonable.

Maxwell had improved on Clausius by showing how to use a velocity distribution for atoms in a gas instead of assuming that they all moved at a single speed. But his derivation of the specific formula for this distribution was, to most physicists, rather abstract and enigmatic. Maxwell reasoned that whatever the distribution formula was, it ought to have two fundamental characteristics. First, it should not care about direction in space. That is, if atoms fill a volume described by three perpendicular axes–the x, y, and z directions of standard geometry–then the distribution of velocities in the x direction ought to be same as the distribution in y and z directions.

Second, and more restrictive, the function should depend only on the overall magnitude of an atom's speed, not on the individual components of its motion in the x, y, and z directions. These two conditions, Maxwell showed, were enough to prove that the distribution must follow the standard Gaussian form, often known as the bell curve of probability theory. Gaussian probabilities turn up frequently in all kinds of statistical analyses. The heights of a random selection of adults, for example, will form approximately a bell curve: people's heights cluster around an average, and the further away from the average one gets, the fewer people there are at that height.

For the case of atoms in a gas, it followed from Maxwell's argument that the square of the atoms' velocities would fall into a Gaussian pattern. It was then easy to calculate the average energy of the atoms (equivalent to the temperature of the gas), the pressure exerted, and so on.

Maxwell's first account of this work appeared in 1860; he followed it up in 1867 with an amplification that tried to provide more justification for his arguments, but this was still an essentially mathematical or logical analysis. Maxwell provided no reason in physics why the atoms should behave this way. His formula for the distribution of velocities was plausible and appealing, but was it correct?

Into this breach stepped Boltzmann. The way that pressure in a column of gas (as in Earth's atmosphere) decreases with height was easily understood from consideration of the gas as a simple fluid under the influence of gravity. If kinetic theory were correct, then the distribution of atomic velocities must also change with height in a way that would give the right answer for the pressure variation. At the same time, a single atom moving upward in a gravitational field is exactly like a ball thrown upward, so that its changing velocity and energy followed directly from consideration of Newtonian mechanics.

Putting all these ingredients together, Boltzmann showed that Maxwell's distribution for atomic velocities gave the correct answer: pressure varied with height in the required manner. But he went further, proposing a general rule. Whatever kind of energy an atom might possess, whether due to linear motion, internal vibration, or gravity, for example, the number of atoms with a specified energy should depend on the energy alone. This amounted to a generalization of Maxwell's formula to include cases where energy was not simply the kinetic energy associated with an atom's velocity. All kinds of energy were equivalent, Boltzmann said, in determining the demographic makeup of a population of atoms.

Boltzmann therefore not only verified Maxwell's arguments but extended them, which is why the resulting Maxwell-Boltzmann distribution is so named. Even so, Boltzmann's ingenious argument was not exactly a proof that the formula was correct. He showed that it was plausible and behaved in the right way in certain well-understood cases, but the argument that it applied with universal generality to any conceivable case was a leap.

Still, Boltzmann provided a large measure of physical justification for what had been until then mostly a mathematical argument. The two contributions of Maxwell and Boltzmann were illustrative of their styles. Maxwell used logical and mathematical reasoning to come up with a formula that perhaps embodied the correct physics; Boltzmann used a direct and intuitive appreciation of physics to show that the mathematical formula indeed seemed to be correct.

Boltzmann was already showing, moreover, an industriousness that was to stay with him for most of his life. By the middle of the following year, still only 25 years old, he had eight scientific publications to his name concerning the behavior of atoms, the physics of electric currents, and general problems in mathematical physics. This early work was, in some ways, a continuation of his schoolboy style: he simply came across problems and solved them, and it was his good fortune to have landed in a university where the new ideas in physics, both atomism and Maxwell's electromagnetism, were welcomed and earnestly debated. It is tempting to believe, in view of what followed, that Boltzmann was somehow predestined to find the main theme of his intellectual life in the development of kinetic theory, but at this early stage it is unclear whether any such destiny was yet calling him. In school, he did well and dutifully all the work that was put before him; in his early research career he followed much the same pattern, guided by the interests of his mentors Stefan and Loschmidt.

It was natural that Boltzmann, an outstanding Austrian student, would go to the University of Vienna, and it was inevitable, once he got there, that he would fall under the influence of the leading physicists around him. What was altogether fortuitous was that those teachers should be as up-to-date and forward thinking as they happened to be. Vienna was distinctly an unusual place in that respect. Maxwell, around this time, became aware not only of Loschmidt's calculation of atomic size and Boltzmann's work in kinetic theory, but also of Stefan's interest in electromagnetism and his experimental investigations of the subject. Mistakenly thinking that Loschmidt was the senior partner in this enterprise, Maxwell wrote to him, "I am very pleased by the outstanding work of your student; in England until now instruction in experimental physics has been much neglected. Sir William Thomson has done the most in this respect, but you are setting us a good example."

These words of Maxwell's were recalled much later by Boltzmann himself in an obituary appreciation describing Stefan's essential role in making 19 Erdbergstrasse into an incubator for a new kind of physics. It is not entirely to Boltzmann's credit that he found room in a eulogy for Stefan to mention a letter from Maxwell praising himself. Throughout his life Boltzmann admired Maxwell and his work above that of any other theoretical physicist, but Boltzmann's later work did not always meet with such warm approval from his British counterpart. These early words of unqualified praise, however, stayed with him. At the time, they must surely have convinced the young, eager, and prolific researcher that he had reason to think of himself as a promising new member of the small but growing society of theoretical physicists.

CHAPTER 3

Dr. Boltzmann of Vienna

The Precocious Genius

THE UNIVERSITY OF HEIDELBERG, in the southern Rhine valley, is the oldest in Germany. Founded in 1386, it has maintained a high reputation for most of the centuries since then, and in the 19th century it had, among other things, a selection of outstanding physicists and mathematicians. It was to Heidelberg that Boltzmann went in the summer of 1870, in his first trip beyond the confines of the Austrian academic world.

There he dropped in on a seminar being conducted by the mathematician Leo Koenigsberger, whose students were struggling to attack a problem he had set. When Koenigsberger asked the class for their ideas, a man he didn't know, "thinner and somewhat older" than his other students, piped up from the back row. The newcomer came down to the front of the lecture room and, directly and clearly, expounded the solution to the problem. He had to compete against a certain amount of giggling from the audience, so amusing was his coarse Austrian accent to the German students.

Koenigsberger asked the stranger who he was. "Dr. Boltzmann of Vienna," said Boltzmann firmly, as if that would suffice, and in fact it did. Koenigsberger knew the name and had heard that interesting work from this young man had already been presented to the Viennese Academy of Sciences. In describing the 26-year-old physicist who had unexpectedly showed up in his class, Koenigsberger used the German word *hager,* which might even be translated as "gaunt" rather than "thin." This was the only occasion when any such word was used to describe Boltzmann; he was an enthusiast for the table, and as a young man soon became plump, then rotund. When he married some years later, his wife, Henriette, took to calling him her "sweet, fat darling."

Boltzmann wanted Koenigsberger's advice on a mathematical problem that had arisen in his recent research. Koenigsberger, impressed by the young man, was only too glad to oblige. But when the two were talking later that afternoon, he asked if Boltzmann had yet gone to see the university's preeminent physicist, Gustav Kirchhoff. Hesitatingly, Boltzmann allowed that he hadn't, and after pressing him a little, Koenigsberger found out why: Boltzmann had discovered an error in Kirchhoff's most recent work, and as much as he wanted to meet the great man, he didn't know whether or how he should approach him. Boltzmann had been in Heidelberg several weeks by this time, failing to overcome his diffidence.

Koenigsberger encouraged the young man to introduce himself to Kirchhoff and to find a way to bring up this delicate matter. Emboldened, Boltzmann left. A few hours later Kirchhoff came to Koenigsberger with an alarming tale to tell. He had been sitting in his office when an abrupt and unruly visitor barged in and, after the barest of introductions, blurted out, "Herr Professor, you have made a mistake!" Kirchhoff, a kindly man but accustomed to rather more deliberate German manners, was taken aback and more than a little suspicious. He wondered briefly if his visitor was deranged in some way. Nevertheless, Boltzmann was able to explain himself, and Kirchhoff realized this awkward young man was right: there was a mistake. In the end, as they got down to discussing their common interests in physics, the conversation proceeded more warmly.

Boltzmann is not known to have had any direct contact with Kirchhoff beyond this single encounter. Years later, delivering an appreciation of the older man's life and work, Boltzmann betrayed no awareness that the oddity of that meeting was in any way a reflection of his own clumsiness. Rather, he explained that although Kirchhoff was by nature "helpful and kindly . . . it took a little effort to warm him up, which was most quickly done through scientific discourse. Then he was enchanting; he stated his opinions without reservation. . . ."

In that first meeting, Boltzmann was caught between two desires. He wanted to spread his name around and meet the outstanding physicists of the day, doing which entailed a certain humility. At the same time, he was eager for the world to know that he had found an error in the work of a senior figure and was punctilious about establishing his priority in such matters.

Through Stefan's influence, Boltzmann was steered into a face-saving subterfuge. By the time he talked to Kirchhoff, he had already prepared a short note for the *Proceedings of the Viennese Academy of Sciences* explaining the other physicist's error and his correction of it, and had written to Stefan, in the latter's capacity as a senior academy member, asking for quick publication so as to ensure his priority. It apparently never crossed Boltzmann's mind that rushing into print to point out someone else's mistake might not be seen as a commendable way to gain scientific kudos, or that there might be a more diplomatic and at the same time more constructive solution. Scientific etiquette would have led Boltzmann to tell Kirchhoff privately of his error, at which Kirchhoff could have written his own note of correction, either with Boltzmann or prominently thanking Boltzmann for his helpful remark. Either way, the science stands corrected, Boltzmann's name is mentioned, and Boltzmann gains an influential ally. Such subtleties, however, invariably eluded Boltzmann.

Nevertheless, he was persuaded to add a small postscript to his published note saying that when he had brought the matter to Kirchhoff's attention, "he let me know he had already noticed it."

Privately, to Stefan, he grumbled that this was not actually true, but who was he to say otherwise?

At this early stage of his career, Stefan's guidance was invaluable to the young physicist. By the time he made his trip to Heidelberg, Boltzmann was no longer based in Vienna, but had recently been appointed a professor of physics at the University of Graz, the same place where Ernst Mach had gone a few years earlier to teach mathematics. Stefan's powerful recommendation had eased his way; he described Boltzmann as a young and promising scientist "who within a short space of time has published a series of works in mathematical physics which give eloquent evidence of his acuity, his assured mathematical knowledge–a man whose extraordinary talent I have been granted the most ample opportunity to admire." Such an endorsement, coming from the head of the Institute of Physics in Austria's capital city, must have opened many a door in the Ministry of Education and Culture, within which all university appointments were decided.

The move to Graz was a good appointment for so young a researcher, but Boltzmann's growing reputation also made him something of a catch for Graz. The university there was old and respectable, but not especially distinguished in the sciences. In 1863, a new medical faculty had been established, and along with it arose a need for improved teaching of science to the incoming medical students. Physics was at the time in the hands of the 62-year-old Karl Hummel, who was distinctly not up to date with recent developments. The university had succeeded in hiring a young physics instructor from Vienna called Viktor von Lang, but he lasted only a year before returning to a better position in Vienna in 1865. The services of Ernst Mach were then engaged for a couple of years, but in 1867, after losing the struggle for the Vienna position that went to Josef Stefan, Mach left to become a professor in Prague, the capital of Bohemia, an important city within the Austro-Hungarian Empire, and home to an old and renowned university.

By this time, the elderly Hummel had been persuaded into retirement, and the University of Graz was able to hire a much livelier head of physics in the person of August Toepler, who had studied in Berlin and was then in his early thirties. Toepler, looking for a junior physicist to act as his assistant, cast his eye across recent graduates from Vienna and, with Stefan's strong encouragement, settled on Boltzmann. With his mother and sister, Boltzmann moved to Graz in September of 1869 to find a tiny and dilapidated physics facility consisting of three small rooms above a lecture hall in what was a converted priest's residence. One room was for a technical assistant, one served as a laboratory, and the third and tiniest was a "chemical kitchen" for experimental preparations. There was little room to do experiments, and little equipment to do them with, but Toepler succeeded in obtaining money to buy new apparatus and hire a lab assistant. He also lent Boltzmann a fur coat–Toepler had been a professor in Riga, on what was then the Baltic coast of Prussia, before coming to Graz–so the younger man could continue his experimental work during the winter in the unheated lab of the Graz physics facility.

In Graz the young Boltzmann had to plan and deliver lectures in elementary physics, which he did with adequate diligence but no marked enthusiasm. With much more eagerness, he applied in the spring of 1870 for permission to spend the summer semester traveling and thus visited Koenigsberger and Kirchhoff in Heidelberg. Here again Stefan's help was invaluable: he used his influence in Vienna to find financial support for Boltzmann's travels.

From Heidelberg Boltzmann went on to Berlin, another great center of German physics, but after arriving on July 5, his plans were waylaid by the outbreak of the Franco-Prussian War, the latest installment of a series of skirmishes and boundary rearrangements that had been plaguing central Europe since the revolutionary year of 1848. Through the middle years of the 19th century, Austria was militarily weak and economically straitened, and its young emperor, Franz-Josef, found himself burdened with the perpetual task of trying to keep his ramshackle realm from falling apart. He had already lost his northern Italian possessions, with the exception of Venetia, and in 1866 had been defeated by

Prussia to the north. Austria and Prussia were the two dominant German powers, but neither wanted to be co-opted into a formal German nation, Austria because it had vast non-German holdings and Prussia because it aimed for preeminence rather than incorporation. In 1864, these two powers had been allies against Denmark in a dispute over the northern states of Schleswig and Holstein, but once the Danes had backed down, Austria and Prussia became uncomfortable co-occupiers of the region. Otto von Bismarck, Prussia's prime minister, maneuvered the far less wily Franz-Josef into a war that Austria lost resoundingly. The Prussian generals wanted to take over Austria completely and occupy Vienna, but now Bismarck put on the brakes: he wanted a reduced but still powerful Austria as a counterweight to the other German states as well as to Italy, France, and Russia. Austria was nonetheless chastened by defeat, and from that point on, Prussia grew stronger and Austria weaker.

The Franco-Prussian War of 1870 was the final move in Bismarck's chess game. This time using a dispute over the succession to the Spanish throne, he goaded France into a conflict in which the southern German states were obliged to take Prussia's side. Austria, as well as Britain and Russia, stayed out of the way. Prussia came out on top, and in the next few years the foundations of modern Germany appeared in the form of a federation with Berlin clearly at the helm. Austria, moreover, was henceforth firmly excluded from the nascent Germany, and was left instead to deal with its own fractious provinces.

Austria's internal weakness had already forced Franz-Josef to accept the institution of the dual monarchy, by which he was emperor of Austria and king of Hungary, presiding over separate parliaments in Vienna and Budapest. Further concessions of power to the Hungarians as well as to the Czechs and others were to follow.

By 1870, the shape of central Europe was essentially fixed for the next few decades. These wars between monarchs and their ministers and generals had for the most part little effect on the lives of ordinary people, and once the fighting was over and a treaty worked out, life quickly resumed its familiar pattern. Just 18 months after his abortive trip to Berlin, Boltzmann was back there again, keen to expand the circle of his scientific contacts and in particular to get acquainted with the rather imperious leader of German physics, Hermann von Helmholtz. The son of a schoolteacher, Helmholtz was fascinated by physics when young, but he studied medicine because he could get financial assistance in return for serving eight years as an army surgeon. He kept up with physics in the meantime through his own efforts, and in 1847, while still in military service, he published a groundbreaking piece of work that set down as no one had done before a systematic mathematical treatment of the conservation of energy. The idea that energy could be neither created nor destroyed was not new, having first been proposed in a recognizably modern form in 1841; the key to seeing the universality of the principle was the recognition that heat was itself a form of energy, not a distinct substance. Rumford's observations in the late 1700s of drill bits boring into cannon metal had hinted that the production of heat was intimately connected to the expenditure of mechanical energy in turning the drill, but it was some decades still before the connection became both unarguable and quantifiable. Helmholtz's work in 1847 in essence tied up all the loose ends and made the principle of energy conservation into the inviolable law it is now known to be.

Thereafter Helmholtz was able to devote his life to research. His interest in music and knowledge of physiology led to him to some notable advances in the science of acoustics and sound perception. His versatility and his personal determination made him into an energetic scientific leader of the German physics community, and by the time he was installed as a leading professor in Berlin, he was well on his way to becoming the "Reichschancellor" of German physics. Michael Pupin, an American physicist of Serbian origin who visited Berlin in the mid-1880s, portrayed Helmholtz as an imposing man, with a large head supported by a muscular neck, but with incongruously small hands and feet and a surprisingly delicate voice. Pupin was introduced to the great man by a subordinate who "bowed before his master as if he wished to touch the ground with his forehead." Helmholtz was formal and punctilious, not easily approached by his students or colleagues.

To Boltzmann, raised in the convivial atmosphere of Stefan's Erdbergstrasse institute, this was all rather strange and forbidding. His admiration for Helmholtz's scientific prowess and his eagerness to make his acquaintance were tempered by the difficulty of doing so. From Berlin in January 1872, Boltzmann wrote to his mother that he had succeeded in having an interesting exchange with Helmholtz, which was especially valuable since Helmholtz is "not so accessible. . . . Although he works near me in the lab, I haven't spoken with him very much."

Helmholtz appeared to be the stereotypical Prussian indeed, immune if not positively hostile to Boltzmann's casual Viennese ways. On one occasion during this visit, Boltzmann's manner earned him a withering look from Helmholtz, which was interpreted to him by one of the junior scientists there: "You are in Berlin now," he was informed.

On the other hand, Helmholtz quickly grasped the aim and importance of Boltzmann's work. While in Berlin, Boltzmann presented some of his ideas on the kinetic theory of gases to a meeting of the German Physical Society and got into a lively discussion with Helmholtz afterward. It didn't appear to Boltzmann that anyone else had much idea of what he was talking about.

Understandably so, perhaps. By this time, Boltzmann was taking kinetic theory into uncharted waters and was close to one of the greatest theoretical achievements of the era of classical physics. Although he had provided, a few years earlier, some physical justification for the correctness of Maxwell's formula for the distribution of atomic velocities in a gas, Boltzmann was keenly aware of what was still missing. There was as yet no argument to say how and why a collection of atoms, banging endlessly into each other, this way and that, should come to obey the Maxwell-Boltzmann formula, or whether, having reached that distribution, it would stay that way indefinitely.

Because they collide so frequently, the speeds and directions of individual atoms are ever changing. An atom that's going faster than the average at one moment might smash into another atom and find itself suddenly moving much slower than the average. Any mathematical formula that purports to describe the overall distribution of atomic velocities in a stable way must clearly be some sort of average. It must describe, at any given moment, the typical number of atoms moving at any given speed, but it cannot hope to be a complete listing of the exact state of motion of every single atom.

Here was the problem that Boltzmann recognized: How is it that atoms in a constant chaos of motion, forever crashing into each other, speeding up, slowing down, changing direction, all in seemingly unpredictable ways, nevertheless maintain an average distribution of velocities that follows a simple, invariable formula–the Maxwell-Boltzmann distribution? How do randomness and unpredictability on the scale of individual atoms give rise to order in bulk?

With the help of Loschmidt's estimate of atomic size, Boltzmann knew that even a small volume of gas must contain trillions upon trillions of atoms. Attempting to follow the motion of every one, keeping track of every collision, every change of speed and direction, was clearly an impossible task. To make any further progress with kinetic theory, Boltzmann had to employ not simply a good deal of mathematical sophistication, but also a certain amount of brute force, along with a powerful belief that somehow an answer must be possible.

Boltzmann's attack on this problem further revealed his capacity to discern the essential physics governing a complex problem and to use that understanding as a means of forging through to the solution. It was one thing, as Maxwell had done, to rely on a sense of mathematical consistency and elegance in order to devise a simple formula for the distribution, one that seemed to be right, appeared to behave in a reasonable way, but that lacked a true foundation in physical theory. Boltzmann, by contrast, was mathematically sophisticated but not necessarily stylish. The important thing was to figure out the answer. Many of his students from his later years recalled a favorite phrase: "Elegance," Boltzmann liked to tell them, "is for the tailor and the shoemaker."

In analyzing the full problem of atomic motions and collisions, Boltzmann's mathematical power gave him the strength to plunge ahead. He did not want to take any short cuts. Picturing atoms as tiny hard spheres, he wanted to apply Newton's laws of mechanics to their behavior and try to figure out from these elementary principles how a vast assembly of atoms, too numerous to be counted, would disport themselves. The problem seemed overwhelmingly–almost ludicrously–intractable, but he was confident it could somehow be solved. He was sure that atoms existed, and sure that they were obedient to the laws of mechanics. Somehow, in nature, hordes of teeming atoms contrived to behave in orderly and predictable ways. The process had to be understandable.

Achieving that understanding, however, demanded every ounce of Boltzmann's intellectual fortitude. He was pushing into a dark and tangled jungle of mathematics, forging painfully and slowly on, sustained only by a belief that somewhere ahead lay a mountain peak from which the whole landscape could be surveyed.

Take a volume of gas and freeze it for an instant in time. Every atom will be caught in some precise state of motion. In principle, Boltzmann saw, he could write down a mathematical catalog of the atoms, allotting to each one its particular speed and direction of motion at that time.

Now unfreeze the gas and let time move on again; the atoms start crashing around and into each other. A moment later, this atom is moving faster, that one is moving slower, and all are moving in different directions. The catalog of atomic speeds and directions has to be drawn up all over again, from one instant to the next.

The daunting task that Boltzmann set himself was to see if he could find a way of tracking how this detailed catalog of atomic motions changes from moment to moment, in order then to see what regularity emerged. He couldn't hope, of course, to track the catalog with absolute fidelity, since that would mean literally keeping an account of every atom and every collision between atoms.

Rather, he would begin with a statistical formulation of the state of the atoms, allowing a completely general specification of how many atoms have velocities in a certain small range and in a certain narrow bundle of directions. Then he would estimate how many collisions between atoms coming at each other with a certain relative speed and at a certain angle would occur in a small interval of time. From that he could calculate how each such class of collisions would alter the motion of the atoms involved. And having done all that, he would finally perform a grand averaging over the whole complicated mess to see how the totality of all possible collisions would change the overall distribution of atomic speeds and velocities.

In taking on this fearsome task, Boltzmann had to combine straightforward and uncontroversial analyses of the mechanics of collisions with much more novel elements of statistical theory. This marked a turning point in the development of theoretical physics and in the understanding of the processes that theoretical physicists were striving to understand. As a discipline of mathematics, the study of probability and statistics had some history, going back to the efforts of Blaise Pascal in the first half of the 17th century to work out the odds governing various dice and card games. But such ideas had remained the province strictly of mathematicians. Laws of probability were all very well for games of chance, but they could not be laws of physics.

The very definition of a law in physics seemed to demand certainty, not probability. Boltzmann himself did not at first appreciate the magnitude of the revolution he was about to instigate. He knew that any particular set of atoms or molecules must have, at any particular time, a certain set of motions; the problem was that no one could possibly know what those motions were with any exactitude. His use of methods from statistical theory was, so he thought, simply a mathematical technique to help him solve a hard problem. It was not that there was anything intrinsically statistical about the motions of the atoms themselves, rather that he had to use statistics to describe atoms because of their enormous number and the almost impossible complexity of their behavior.

But even in this limited sense, statistics and probability were foreign subjects for most physicists of the time. The idea of a mathematical function representing not the actual specific state of a lot of atoms but rather the likelihood of their being in this or that state was a strange and slippery notion. The idea, furthermore, of trying to understand the effect of statistically averaged collisions on this statistical description was, to many physicists of the time, outlandish to the point of incomprehensibility.

Nevertheless, Boltzmann pressed ahead. In a 100-page paper published in Vienna in 1872, with the unrevealing title "Further Studies of the Thermal Equilibrium of Gas Molecules," Boltzmann laid out in minute detail his analysis of atomic velocity distributions, and established a series of profoundly important results. The bulk of the paper was taken up with the establishment of an equation–Boltzmann's transport equation, as it became known–which embodied the variation of the distribution of atomic velocities due to collisions. Through pages of laborious but purposeful calculation, Boltzmann established how a typical or average set of collisions would transform the distribution. The only gross simplification he could allow himself was to assume that the atoms are moving in random directions, an entirely reasonable assumption since the volume of gas in question is not, as a whole, moving anywhere. He established, in the end, a surprisingly simple differential equation.

That done, he had to work out what might be the solutions to this equation. A general solution was impossible; that would be tantamount to understanding in complete and microscopic detail the behavior of any random set of atomic velocities. But Boltzmann was interested in a more specific case: thermal equilibrium, as it is called. Physicists had long understood that a volume of gas at a fixed temperature will exert a predictable pressure, and that if the gas is left alone it will remain in that fixed state indefinitely. Compress the gas and its temperature and pressure will go up in an orderly manner, adjusting to the new volume by establishing itself in a new equilibrium state.

The defining characteristic of equilibrium, from Boltzmann's new perspective, was that even though individual atomic velocities are constantly changing, the overall distribution is not. As many atoms are boosted to higher velocities as are demoted to lower ones. Boltzmann sought out a solution to his new equation that would correspond to such a state. It was, after all the preceding work, rather easy to find. There was one and only one unchanging or "stationary" solution and it was, to Boltzmann's immense satisfaction, none other than the Maxwell-Boltzmann formula.

Boltzmann had now proved that the distribution he and Maxwell had arrived at through a mixture of guesswork and arguments from plausibility was not just the right one but indeed the only possible one. Finally this was a proof, starting from nothing but Newton's laws for the collision of atoms, that a state of thermal equilibrium must correspond to a Maxwell-Boltzmann velocity distribution, and that the Maxwell-Boltzmann velocity distribution was the only one corresponding to thermal equilibrium.

Boltzmann's great work of 1872 signified the arrival of a true genius of physics. His earlier work had been notable but, as is often the case in science, it was work that any one of a number of leading physicists could equally well have produced. In his 1872 publication, however, Boltzmann bulldozed a path through thickets of reasoning and intricacies of mathematics that no one else would even have dared to tackle. He succeeding in extracting a powerful equation and a simple answer where, at first sight, it was hard to see how to make any progress at all.

That, at least, is how Boltzmann's achievement appears now. At the time, few physicists were capable of understanding his aims and methods, and fewer still had the tenacity to work through his pages of calculation. In his scientific work, as in his conversation and his personal letters, Boltzmann set down his ideas more or less as they came to him. It was not his habit to refine his writings in order to make the flow of logic clearer to the uninitiated. It apparently never occurred to him that this would be a useful thing to do, either for his readers or, indirectly, for his own benefit, in that his readers might more easily see what he was about.

Clausius, the acknowledged originator of modern kinetic theory, lacked the mathematical acumen to follow his younger colleague. Even after Maxwell had introduced the idea of a distribution of atomic velocities, Clausius never abandoned his earlier practice of treating all the molecules in a gas as if they had the same average speed. If that step was beyond him, surely Clausius could not follow Boltzmann's further elaboration of analyzing how the distribution itself changed with time. And no one else in the German physics world had anything close to Boltzmann's intensity of interest in kinetic theory.

Across the English channel, Boltzmann's work found a more alert audience. Sir William Thomson, late in 1875, found himself musing on the Austrian's arguments during a train journey, and afterward jotted a note to a colleague: "it is very important . . . the more I thought of it yesterday in the train, the surer I felt of its truth." Maxwell, too, was following Boltzmann's achievements, but his thoughts on kinetic theory had begun to take a slightly different tack, and he found himself unable to fully accept all of Boltzmann's conclusions. It appears that even he, who among all physicists was capable of understanding the argument leading up to the new theorem, never pushed himself to make the effort. Word began to get around that Boltzmann had done something remarkable, but hardly anyone seemed to understand quite what it was, and it was only a few years later, when objections and counterarguments surfaced, that even a moderate number of physicists set themselves the task of trying to follow Boltzmann's derivations.

Boltzmann had in fact achieved not only a profound result in physics, but one that brought into the world a new style of reasoning. He had used an essentially statistical analysis to establish an absolute truth, the correctness of the Maxwell-Boltzmann formula. But the revolutionary nature of this argument was not fully apparent at the time, even to its author. Boltzmann believed he had solved the essential problem of kinetic theory, and, finding little reaction from his peers, wrote nothing more on the subject for a number of years.

Boltzmann's life at this time was simple and comfortable. He worked at his physics, and he lived in modest accommodations with his mother and his sister Hedwig, four years his junior. He gave lectures, favoring applied mathematics and the more mathematical areas of physics, especially, of course, the mechanical theory of heat to which he had already contributed so much. August Toepler provided money for some experiments, improved the state of the physics facility in Graz, and generally took care of any administrative matters. Although he was fond of weekend rambles in the countryside with his mother and sister and with some of his university colleagues, Boltzmann was a young man whose energies and thoughts were rarely distracted from physics.

One distraction had shown up, however, in the person of Henriette von Aigentler, a young student at the teacher training college in Graz. Henriette knew Boltzmann's sister through the local teacher training institute, and in May 1873 she met Boltzmann on a school outing. Fräulein von Aigentler, then 19 years old, was an intelligent and determined woman. The year before she met Boltzmann, she had decided she wanted to sit in on science courses at the university, even though women at that time were unable to take degrees. It was generally the opinion that the presence of women in class would distract the male students, and that in any case the female intellect was insufficiently rational for the appreciation of chemistry, mathematics, and physics. This was the attitude of Boltzmann's boss, August Toepler, who refused to let Henriette into the classroom. But she did not give in. She obtained references from lecturers whose classes she had visited testifying to her quiet and respectful demeanor, thus managing to win the approval of the university's officials. In the winter semester of 1872 she started attending science classes, although she had to fight constantly to obtain renewed permission to sit in on lectures.

What contact there may have been between Henriette and Boltzmann during the summer of 1873 is unclear. In August of that year, however, Boltzmann left Graz to return to Vienna, where he had been appointed a junior professor of mathematics. Teaching mathematics was not his ideal choice, but the attractions of Vienna overcame any doubts on his part. Nor, evidently, had the attentions of Henriette von Aigentler impinged on him enough to cause him any hesitation in deciding to leave behind Graz and its middling university.

But Henriette did not abandon the connection she had made. In October she wrote her first letter to Boltzmann, asking for advice on her studies. She apologized for burdening him with her questions, but explained that since her father was no longer alive and her mother knew nothing of these matters, she had no one else to turn to; in any case, she added, Boltzmann's sister had assured her that he would treat her requests sympathetically. Boltzmann's replies to Henriette's first letters do not survive, but she was evidently thankful and continued to write.

Her third letter, written in March of the following year, takes a serious turn. She let Boltzmann know that her mother had taken sick the previous Christmas and had died on December 30. Henriette, just turned 20, and the youngest of three daughters, was now an orphan. But she was continuing to study and again asked Boltzmann for advice and help. Fortunately she had connections in Graz, since her father had been a civil servant of some standing, and she ended up living with the family of the mayor of Graz. Still, her future must have seemed unsettled to her, and she began to turn to Boltzmann with increasing frequency and insistence. She wrote to him in April 1874 with news of her studies, and then again in June to congratulate Boltzmann on his election as a corresponding member of the Viennese Academy of Sciences, which she had read about in the newspaper.

Still there is no record of Boltzmann's response. If he was aware of what it meant to be receiving increasingly insistent letters from an attractive young woman 10 years his junior, he showed no sign of it, or perhaps simply had no idea what was expected from his side of the exchange. But Henriette showed as much persistence in pursuing the physicist as she had already shown in obtaining permission to attend physics and other classes. In November 1874, Boltzmann wrote to her from Vienna–the first surviving record of his side of the correspondence. It is a short but sympathetic note, commiserating with Henriette over yet another death in her family, this time of her married sister.

Now Henriette was able to make her next move. She wrote again in December, and after giving more news of her studies made a request: "there is something else close to my heart. I have wanted to ask you about it for a long time, but didn't dare. I would so much like a memento of you, namely your photograph. Really, can you send me one? It would be an object of my most sincere admiration as long as I live. Hoping for my wish to be granted. . . ."

Still Boltzmann was slow on the uptake. When he failed to comply promptly, Henriette sent another earnest and entreating letter, repeating in no uncertain terms her request for a photograph: "it's admittedly a presumptuous request, but if you knew how much I wanted it, perhaps you would not take so long." Finally Boltzmann sent a picture, with an apology that he had been unwell for a time. Henriette immediately sent a photograph of herself in return, and wrote shortly after to say how often she looked at her picture of him. She now kept up a steady stream of chatty letters to Boltzmann, and to his often tardy or perfunctory replies she was quick to respond with pleading and worried notes, hoping he is well, and not upset by her insistence.

It was now the summer of 1875. In July Boltzmann wrote to say he would be visiting Graz in September to attend a scientific conference and hoped they could meet. Back in Vienna again after the conference, he wrote to Henriette on September 27, asking her to marry him. His letter of proposal was serious and considerate, if not exactly passionate. He began by declaring that she had made a deep impression on him from their first meeting, and that as he got to know her better, he found in her those qualities that seemed to him most apt to underpin "a lasting sympathy between us." He goes on: "it seems to me that lasting love cannot exist if a wife has no understanding, no enthusiasm for her husband's striving, and is merely his housekeeper, not a comrade in a shared endeavor. Understand by this my confession that I love you."

Henriette's response has been lost, but it was rapid and positive.

Now they wrote to each other constantly. Between the proposal and their marriage the following July, over 100 letters and a handful of postcards flew between Graz and Vienna. The correspondence is voluminous but, as to the personalities of the participants, oddly unrevealing. Along with gossip about their daily lives, there are extravagant protestations of love and expressions of anguish that the two won't see each other again for a number of days. Both writers draw little hearts at the end of their letters: "these hearts bring you my hottest kisses." But from neither party is there introspection or soul-searching about the nature of their love. With the proposition of marriage offered and accepted, both Boltzmann and Henriette seem mainly concerned with sorting out the immediate logistics of their lives, not their grandest ambitions or hopes.

While all this was developing, Boltzmann's position in Vienna was not turning out as he might have hoped. The Institute of Physics was no longer in the Erdbergstrasse house Boltzmann so fondly remembered, but had moved to a new building, a converted apartment house in Türkenstrasse. He in any case had little time for physics, because of his duties as a lecturer in mathematics, which, as he had guessed, turned out not to suit him very much. He was mathematically adept–as his fundamental proof of the Maxwell-Boltzmann distribution had recently and amply demonstrated–but he was by no means a mathematician. The distinction may need some explanation. Nowadays it can easily seem that theoretical physics has become as much mathematics as physics. A page from a physics journal may look as abstruse and intimidating to the unenlightened observer as a page from a journal of mathematics. But there is a broad difference. Physicists, for the most part, take up mathematical ideas that have been developed by others, and adapt them to make physical models. They do not generally invent the mathematics they use.

Newton is a singular exception to this rule. In order to figure out how planets would revolve around the sun if controlled by an inverse square law of gravity, he had to come up with a new kind of mathematics called the calculus, and for this he is revered as a great mathematician as well as a great physicist. But he is really the only such person to be so regarded. Einstein, for example, introduced the mathematics of curved spaces into physics, but he took what he needed from mathematicians who had developed non-Euclidean geometry during the second half of the 19th century.

Theoretical physicists, even the great ones, tend to take mathematics as a set of tools and don't spend too much time worrying about where mathematics comes from, or why and how it all fits together. Those abstract questions are the mathematician's responsibility.

More mundanely, as Boltzmann himself well realized, there were large areas of mathematics that he knew next to nothing about. He was well able to teach a class on differential equations or on the theory of statistics and probability that he had made such good use of. But he was in no position to teach a class on elementary number theory, for example, which concerns itself with such things as the properties of prime numbers and the difference between the rationals and the transcendentals. Boltzmann had from the outset had some hesitation about his suitability for the mathematics position that had opened up in Vienna.

Still, Graz had a respectable but not particularly distinguished university, while Vienna was Vienna, the apex of the Austrian academic world. Boltzmann's teacher and mentor Josef Stefan was keen to bring his brilliant young student back to the capital, and the elderly mathematics professor whose retirement made the new position available himself expressed enthusiasm for Boltzmann's mathematical abilities.

Boltzmann's doubts about his new job soon proved well-founded. Fortunately, his duties as a young professor of mathematics were only nebulously described, a circumstance he took advantage of by teaching classes in applied mathematics with, naturally, a particular attention to the mechanical theory of heat and the kinetic theory of gases. He cajoled the university into giving him some money to continue a little experimental work in physics, and at the same time traveled frequently back to Graz to work on experiments there, with Toepler, as well as to visit Henriette. During his sojourn in Vienna as a mathematics professor, following his monumental proof of the Maxwell-Boltzmann distribution, he published little in mathematical physics and devoted most of his energies to his interest in experimental physics, particularly in the measurement of electrical behavior to test its conformity to Maxwell's electromagnetic theory. He continued to publish at a prodigious rate–some dozen scientific papers in three years–but it was only toward the end of this time that theoretical interests began once again to engage his attention.

In the meantime, the young physicist took the opportunity to learn something of the game of academic career advancement. In early 1875, the prestigious Polytechnic Institute in Zurich, Switzerland, made him an attractive offer. Boltzmann was interested, but despite his reservations about his duties in Vienna, he didn't really want to leave. Nevertheless, he dangled the Zurich offer before the Austrian ministry and was able to obtain for himself a substantial pay raise, more money to do physics, and on top of that a written undertaking that should the university hire another lecturer in mathematics, he would be free to shift his interests more overtly toward research and teaching in physics–all this while remaining, nevertheless, a professor of mathematics.

He even tried the same trick again later that same year, when he was approached by the University of Freiburg in southern Germany. Academically, Freiburg could not compare to either Vienna or Zurich. On the other hand, Boltzmann could be director of the physics institute there, and, as Henriette wrote to Boltzmann, Freiburg might be a cheaper place to live and was in an attractive setting. Its being a small town, she added, "is an advantage for our personal life, because the conveniences of a large city have no value for us."

In the end, however, Freiburg could not come up with an offer that Boltzmann (taking Henriette's opinions into account) found sufficiently generous, and the Austrian ministry, having given so much to Boltzmann earlier that year, was not inclined to open its pockets again. He remained in Vienna.

But not happily. The task of teaching mathematics soon became, as he had suspected it might, more work than he cared for. He kept up a regular correspondence with Toepler, back in Graz, but his letters generally contained little more than chit-chat about physics and gossip about people they both knew. His occasional letters to Helmholtz, on the other hand, took on at times a surprisingly confessional tone, in the light of his admitted difficulty in talking to the senior physicist during his few weeks in Berlin. In between requests for technical information and advice, Boltzmann let on that he didn't find lecturing in mathematics all that congenial, that his teaching was taking away from the time he could spend on physics, and that he would rather be a professor of physics except that no suitable position was then available. In one letter, he even told Helmholtz that his salary was barely adequate for life in "so tremendously expensive" a city, and that there were times when he wanted to live "not as a physicist but a little as an ordinary human being."

These seemingly artless revelations may have sprung from an ulterior motive. Helmholtz was an enormously influential figure, Berlin was an important center of science, and Boltzmann may have figured it would do no harm to keep Helmholtz apprised of his unhappiness in Vienna in case anything should turn up. Unfortunately, Helmholtz's replies to Boltzmann have not survived; Boltzmann later expressed regret that he had not kept them, but as things turned out there came a time when he had reason not to want these reminders around him.

It was not too long, in any case, before a chance to get away from Vienna presented itself. Back in Graz, his mentor Toepler was wearying of the task of running the physics department, and on top of that had broken a rib falling down an elevator shaft in the dilapidated old physics building there. At the same time he was being tempted by an attractive offer from the university in Dresden, which, after some vascillation, he accepted.

This opened up a senior position in Graz, for which Boltzmann was clearly a leading contender. It was explicitly a job for a physicist, and strictly speaking, for an experimental physicist. Whereas a ministerial document recommending Boltzmann's appointment in Vienna three years earlier had emphasized his mathematical astuteness, a similar recommendation for the Graz position now emphasized how productive he had been in the laboratory. This was all true, to an extent, although his experimental findings never achieved the greatness of his innovations in theory. By this time, moreover (he was now in his early thirties), Boltzmann's eyesight, poor from birth, was failing further. It became increasingly hard for him to do experiments, and as the years went by he was forced to rely increasingly, and then entirely, on the assistance of others in the lab.

Although Boltzmann had strong backing for the Graz position, an effort arose to bring Ernst Mach back from Prague, where he had now been for almost 10 years. Mach had, as it happened, married a young woman originally from Graz, while Boltzmann's marriage to Henriette, set for July 17, gave him a sentimental reason to return to Graz as well.

Henriette found herself in a position to help her future husband, or at least deal in rumors and gossip. Because she lived in the household of Herr Kienzl, the mayor of Graz, she had connections with important people in the town and the university. She also knew Karl von Stremayr, a civil servant now with the Ministry for Education and Culture who had earlier worked with her father.

Not only that, she had a line of communication to the Mach camp, such as it was. The son of the mayor of Graz was one Wilhelm Kienzl, who during his lifetime became a well-known composer. He was a noted Wagnerian and his most famous work, an opera called *Der Evangelimann* (loosely, *The Preacher*) drew large audiences at its Berlin premiere in 1895. As a young man Kienzl was interested in both physics and music and had gone to study in Prague, where he took lectures from Mach. Mach's candid assessment of Kienzl's showing as a physics student solidified the young man's determination to become a musician. In his autobiography, the composer presented a brief picture of Boltzmann at about this time, describing him as "a strong, heavy-browed man, very shortsighted and bespectacled on that account, with tightly curled brown hair and a full beard framing a broad, flushed face, always somewhat stooped in posture."

In early June, only a few weeks before the wedding, Henriette was able to relay to Boltzmann a bit of news from Frau Kienzl, which was that she had spoken directly to Mach to let him know that Boltzmann was interested in the Graz position. Mach supposedly said that in that case, although he too would have liked the job, "if he had to make a proposal he would propose you as first choice." Henriette also used her connection to Stremayr to intimate that her fiancé's chest was inclined to be weak, and that the mountain air around Graz would be excellent for his health. She also reported that an official in Graz had told her that Mach is not likely to get the job because he is "industrious but no genius."

The significance, if any, of these minor conspiracies is impossible to assess. In the very last week before their wedding day, Boltzmann's agitation over the uncertainty boiled over. He began to worry that if, as planned, they were to go away to Switzerland on their honeymoon, and the decision were made in his absence, he might lose not only the Graz job but also some advantages of his position in Vienna. Abruptly, he proposed that after they get married, they should live in Vienna while things were worked out.

Boltzmann's suggestion upset Henriette. "That all the beautiful hopes for our honeymoon should come to nothing!" she exclaimed. Rather than fall in with his new plan, she told him she would prefer to postpone the wedding altogether. Hastily and apologetically Boltzmann wrote back, saying that he had now had further words with someone in the ministry, who told him that a decision over Graz would not be made until the second half of August, so there was no point in postponing the wedding. "There is really nothing more I can do in Vienna," he said.

So, five days later, on July 17, 1876, they were married in Graz as they had intended all along and went off to Switzerland for a honeymoon. Despite all his concerns, Boltzmann duly became professor of physics and director of the physics department in Graz. He was then 32 years old, and Henriette 22. Boltzmann had at first suggested to his fiancée that they should all live together, the newlyweds along with his mother and sister, but Henriette, as his wife, decided otherwise. The married couple found lodging for Boltzmann's mother and sister elsewhere, while they at first lived in accommodations provided by the university.

Children soon began to arrive. Their first son and daughter, named Ludwig and Henriette after the parents, were born in 1878 and 1880. A second son, Arthur, came in 1881 and another daughter, Ida, in 1884. Their last child, Elsa, was born some years later, in 1891, after the Boltzmanns had left Graz. Despite her earlier strenuous attempts to study science, and despite Boltzmann's insistence that his wife should be a comrade-in-arms and not a housewife, Henriette abandoned her academic plans, took cooking lessons from Frau Kienzl, and embarked on a life of seemingly typical domesticity. The Boltzmanns built themselves a house on farmland a few miles northeast of Graz, on the slopes of a mountain with splendid views over the surrounding country. Boltzmann doted on his children. He took them on walks in the countryside around Graz, teaching them the plants and flowers of the region. (A colleague at Graz, a professor of botany, said later how impressed he was with Boltzmann's knowledge.) His devotion to his children took him on one occasion to an impractical extreme: he decided his daughters needed fresh milk, so he bought a cow at the local market and walked it home through the streets of Graz. But then he had to consult a zoology professor at the university to find out what one should feed a cow, and how one should arrange things so that it produced milk.

His years in Graz were the most productive of Boltzmann's career. At just about the time he left Vienna, his earlier work in kinetic theory finally began to gain some broader attention–albeit critical attention–and as he defended his ideas, Boltzmann elaborated them further. But at the same time he continued to work in the laboratory and to cast his gaze at other issues in mathematical physics. In one notable achievement, he was able to repay the debt he owed to his first teacher, Josef Stefan, who had instilled in him an interest in electromagnetic phenomena in general and Maxwell's theory in particular.

Stefan, in 1879, had established experimentally that electromagnetic radiation in thermal equilibrium had an intrinsic energy proportional to the fourth power of the temperature. In 1884, Boltzmann used Maxwell's theory in combination with his sophisticated understanding of heat and energy to propound a theoretical explanation of this relation, and also to prove that the radiation would exert a pressure proportional to the cube of its temperature. This result, in which Boltzmann established a fundamental connection between radiation theory and thermodynamics, is now known as the Stefan-Boltzmann law.

Once he had his own institute in Graz, Boltzmann tried to reproduce the happy atmosphere he had enjoyed in his younger days in Erdbergstrasse, and to some extent he succeeded. As his name became known across Europe, a few students came to Graz specifically to study with him. Walther Nernst, who was to win the 1920 Nobel Prize in chemistry, visited in 1885 and was at first disappointed to find Boltzmann so busy with his introductory lectures in mathematics and physics that he had little time to spare. But then Boltzmann suggested an experimental investigation that might be worth doing, and once Nernst was embarked on this advanced project, he found Boltzmann happy to spend hours discussing its finer points. Nernst remembered a "well-organized institute, in which teachers and researchers worked together with students in an exemplary fashion."

A similarly warm recollection came from Svante Arrhenius, who came to Graz from Sweden in 1887 and who was a Nobel Prize winner for chemistry in 1903. Like Nernst, he recalled Boltzmann being willing to discuss and debate science at great length with advanced students, but admitted that only a few enjoyed such close contacts. In the end, Boltzmann did not establish a school of his own with anything like the influence of Stefan's Erdbergstrasse institute. In part, the students at Graz were not of such high quality or ambition as those in Vienna, but in part it seemed that Boltzmann would only engage directly those few promising young scholars who made the extra effort to seek him out.

Academic and civil distinctions came his way. He was made "government councilor," an honorific of the Habsburg court, and some years later "court councilor." He became a full member of the Viennese Academy of Sciences, and foreign academies honored him with membership. But at the same time Boltzmann would often feel isolated in Graz, away from the great academic centers of Europe. Besides Nernst and Arrhenius, he had little influence on young researchers destined for greatness. For several years after settling in Graz with his new wife, he hardly traveled at all, even to Vienna, and although he kept up a mostly personal correspondence with Toepler, he had only irregular contact with more notable researchers in Austria and Germany.

Even in the small world of Graz itself he had reclusive tendencies. His academic position as well as Henriette's childhood in the Kienzl household gave the couple an automatic status in social circles, but they seemed to have little use for such things. One faculty member recalled that "with the physics genius Ludwig Boltzmann, who at that time already stood at the height of his glory, I like the rest of my colleagues had little personal contact, on account of the withdrawn way this shy eccentric lived." Wilhelm Kienzl, the composer, offered a similar assessment. Depicting Boltzmann rather as Boltzmann depicted Loschmidt, Kienzl called him "the prototypical unworldly scholar, living wholly in the realm of his science and his groundbreaking research. . . . He commanded a broad range of general knowledge, which had no impact whatever on the manifestly childish naivete of his nature, as one often finds with those whose focussed minds move in higher spheres."

For over a decade Boltzmann lived and worked in Graz, his scientific output steady, his name known increasingly throughout the scientific world, but maintaining little direct contact with his scientific colleagues elsewhere, and in person living with his family almost in solitude.

CHAPTER 4

Irreversible Changes

The Enigma of Entropy

THE ONLY MAN BESIDES BOLTZMANN to grasp the growing importance of statistics and probability in physics was James Clerk Maxwell in England. At the age of 19 he had come across a book by the Belgian mathematician Quetelet in which statistical analysis–for example, of the range of heights of a group of soldiers–was expounded in a distinctly modern way. Quetelet also hinted that statistics could usefully adopt the rigorous methods employed in physics, while Maxwell in turn saw that physics might benefit from the introduction of statistics. "The true Logic for this world is the Calculus of Probabilities," he observed in a letter; "This branch of Math., which is generally thought to favour gambling, dicing, and wagering, and therefore highly immoral, is the only 'Mathematics for Practical Men.'"

He was quick to capitalize on this insight, publishing in 1859 a groundbreaking analysis in which he used a new style of argument to prove that the rings of Saturn must be composed of numerous tiny particles. Galileo, with his first telescope, was surprised to find that Saturn appeared to have "handles"; Christiaan Huygens, using a better instrument, had in 1656 concluded that a ring encircled the planet, and later observations revealed that the ring was multiple, consisting of bands and gaps. By the middle of the 19th century, the nature of these rings remained unknown, and in 1855 the structure and stability of Saturn's rings was set as the topic for Cambridge University's Adams Prize. Maxwell, adept at mathematics and confident of his knowledge in mechanics, tackled the problem.

It proved more difficult than he had expected, and he wrestled with the analysis for two or three years before he had sorted it out to his satisfaction. Solid rings were impossible, as others had already concluded; the planet's gravity would try to make different parts of the ring rotate at different speeds, creating stresses that no physical material could withstand. Maxwell tried instead fluid rings, and what he called "dusky" rings composed of countless tiny particles, like dust grains. Modeling the latter kind of structure demanded a novel mathematical technique. He could not literally track the actual motion of every such particle; instead, he set up an essentially statistical description of the rings, which allowed for prescribed numbers of particles to follow certain classes of orbit, just as a population census might divide up people according to age, weight, and height classes.

Applying Newtonian mechanics to this dusky ring, Maxwell showed that the particles could move in collective modes corresponding to waves in the ring's density. Only if these waves remained limited in magnitude would the ring itself remain stable, and this entailed certain conditions on the size and number of particles in the ring. He thus proved that Saturn's rings could exist indefinitely, retaining their shape and density, if they were composed of particles of a suitable size. For this demonstration Maxwell won the Adams Prize.

With this considerable achievement in hand, Maxwell was ideally placed to apply the same sort of analysis to gases, in which, the atomists now asserted, broad properties such as pressure, temperature, and so on were to be understood as the gross manifestations of all the tiny and incalculable motions of countless atoms.

Maxwell's proposed distribution of the velocities of atoms in a gas was, in mathematical terms, a straightforward variation on his description of the particles in Saturn's rings. And Maxwell's investigation of the motions and stability of the rings made it an obvious next step for him to think similarly about the stability of atomic velocity distributions. In 1866, he published a long analysis, "On the Dynamical Theory of Gases," setting out everything he had learned about the kinetic model of gases, showing how to obtain from the underlying distribution of velocities all manner of physical properties of the gas itself.

Boltzmann was well aware of Maxwell's 1866 paper. Indeed, on one occasion he flew off into a romantic fancy about its brilliance, rhapsodizing over Maxwell's ability to orchestrate a mathematical argument with symphonic coherence. "First the variations in velocity develop majestically, then the equations of state enter on one side, the equations of motion on the other; ever higher surges the chaos of formulas. Suddenly, four words sound out: 'Put N = 5.' The evil demon V vanishes, just as in music a disruptive figure in the bass abruptly falls silent . . ." Even in physics and mathematics, Boltzmann could find the melodrama he so much admired in music and the theater.

But Boltzmann's sometimes extravagant passion also gave him the courage of his convictions. It was he, not Maxwell, who pushed kinetic theory further. Maxwell hesitated; he saw a problem looming that did not become clear to Boltzmann until some years later. Aware of the work of Loschmidt, Stefan, and Boltzmann, Maxwell had written praising the efforts of the Viennese. But just a few years later, in a letter of December 1873 to the Scottish physicist P. G. Tait, whom he knew from his school days in Edinburgh, Maxwell was mocking the continentals. It is "rare sport," he said, to see "those learned Germans" tangled in confusion.

The cause of Maxwell's aloof amusement was ostensibly another priority dispute that Boltzmann had gotten himself into, this time involving Clausius, the man who had first explained how atomic motion manifested itself as heat. But there were deeper issues here that would dog Boltzmann for many years, and in part cause Maxwell to turn his attentions away from the blossoming kinetic theory of gases. The looming problem sprang from a seemingly banal fact: heat always flows from high temperatures to low, so that anything hot inevitably cools down of its own accord. But why is this so, and why doesn't the reverse ever happen?

In 1865, the same year that Loschmidt had estimated the size of molecules in air, Clausius published an influential work clarifying some formerly elusive notions about the nature of heat, energy, and mechanical work and in the process coining a new word: entropy. The word arose in the context of what is now called the second law of thermodynamics. The first law is the rule that Helmholtz had done so much to enunciate–the law of conservation of energy.

The second law of thermodynamics, like the first, existed in rudimentary form before it found a precise formulation. In 1824, a French engineer named Sadi Carnot came up with a profound but rather mystifying analysis of the efficiency of steam engines. At that time tinkerers and inventors of all sorts were trying to make better steam engines, mainly through inspired guesswork, since there was no theory of steam engines to guide them. Into this breach stepped Carnot, who imagined an idealized engine in which a steam-filled cylinder expands, pushing on a piston and performing mechanical work, then cools and returns to its starting position. By thinking of the transformation of energy and heat involved in this complete cycle of activity, Carnot proved there was a maximum amount of work that such an engine could perform, depending only on the high and low temperatures between which the device cycles.

Carnot's argument, or variations on it, applies very generally. It is why, for example, you can't cool your house down by leaving the refrigerator door open: the fridge uses energy to stay cool inside, but only at the expense of expelling more heat into its surroundings than it removes from its interior. It was thought at first that Carnot's principle was in some way a consequence of the conservation of energy, but investigations over the next several decades, particularly by William Thomson and William Rankine in Britain and by Clausius in Germany, showed that there was a second independent principle at work. These efforts generated the science now called thermodynamics–literally, the dynamics of heat.

Thomson and Clausius particularly used Carnot's insight to understand the nature of thermodynamic changes. In an idealized, isolated system, the total amount of energy must remain constant; that was the first law of thermodynamics. But within that system, energy could change from one form to another and back again. Physicists distinguished two types of changes: reversible ones, in which the system could be exactly restored to its starting point, and irreversible ones, in which it could not–not, that is, without the application of further external energy. In reversible changes, something stayed the same; in irreversible changes, it did not.

That something, Clausius said in 1865, was entropy. In reversible changes entropy remained constant, but in irreversible changes it grew. Entropy in an isolated system can never decrease, which is why irreversible changes are indeed irreversible. Once the entropy has increased, it can't fall back to its former level. By the same reasoning, entropy in any isolated system would tend to increase until it reached its maximum possible value. That state of maximum entropy, Clausius said, was the state of perfect thermal equilibrium. The rule that entropy can never decrease, only increase or stay the same, was a new physical principle: the second law of thermodynamics.

Whereas heat and energy are physical quantities that can be fairly directly grasped, entropy has a more abstract character. It represents a sort of potential energy: mechanical work can be extracted from a system as long as there is room for entropy to increase, but in a uniform volume of gas in thermal equilibrium, entropy has attained its maximum possible value, and no further work can be obtained.

Clausius defined entropy in terms of the heat going in or out of a system and the temperature at which such exchanges are happening. Immediately, enthusiasts for the kinetic theory of heat wanted to understand entropy in terms of the underlying atomic constitution of a gas rather than its overt bulk properties. The temperature and pressure of a gas were simply related to the average kinetic energy of its constituent atoms; that was now straightforward. But what was the kinetic definition of entropy? What quality or average property of moving atoms corresponded to this newly minted thermodynamic quantity?

In 1866, the 22-year-old Boltzmann had published an attempt to answer this question–a very sketchy and preliminary attempt, relying on some restrictive and in truth unrealistic restrictions on the way atoms could move. This was the first attempt on a problem that was to keep Boltzmann busy, one way or another, all his life. It was not, however, a great piece of work, and since Boltzmann's name was unknown at the time, it went largely unnoticed. A few years later, Clausius hit on a similar idea and in 1871 published a brief note saying more or less the same thing. Clausius being who he was, his argument gained a little attention.

Most notably, it drew some attention from Boltzmann, by then in his first appointment as physics professor in Graz. He submitted a lengthy, not to say long-winded, memo to the Viennese Academy of Sciences, in which he reprinted several pages' worth of his 1866 paper and concluded, in case any reader did not see what he was getting at, "I think I have established my priority." That was heavy-handed enough, but Boltzmann went further: "Finally I wish to express my pleasure that an authority such as Dr. Clausius contributes to the dissemination of the ideas contained in my papers on the mechanical theory of heat."

Subtleties of phrasing were never Boltzmann's strong point. A colleague once commented that "style is the man, as a Frenchman has said. . . . Boltzmann wrote good, flowing German, with occasional Austrianisms, but he did not turn his sentences. Everything came out quite unaffectedly, just as it came into his head."

Clausius was presumably not pleased at being thanked by someone he didn't know for acting as the messenger for work he knew nothing about. Still, he published a short, mild reply acknowledging that Boltzmann had indeed had the idea first and apologizing that he had not been sufficiently conversant with the scientific literature to be aware of the younger man's work. But he finished by saying that he thought his result a little more general than Boltzmann's.

Boltzmann might have been inclined to snipe back, but as 1871 progressed he became caught up in the ideas that would power his monumental work of 1872; he saw, in fact, that he could solve this fundamental problem in its entirety. His analysis of atomic collisions, derivation of the transport equation, and proof that the Maxwell distribution was the only possible distribution corresponding to thermal equilibrium were (to use a symphonic analogy Boltzmann might have liked) merely the overture to what was the great theme of his 1872 analysis. The culmination was something Boltzmann called his minimum theorem, a result that some years later became known as the H-theorem when an English physicist apparently misread a German script upper-case *E* in one of Boltzmann's papers for an *H*.

What came to be called H, at any rate, was a numerical quantity defined in terms of the velocity distribution of the atoms, whatever form that might take. For any collection of atoms, moving at any assortment of speeds, the value of H came from a formula that Boltzmann devised.

The significance of H was twofold. First, when the atoms fell into a Maxwell-Boltzmann distribution, H assumed its minimum possible value. Second, Boltzmann argued, a collection of atoms whose H-value was greater than this minimum would, through the effect of collisions, transform its distribution of velocities in such a way as to decrease H, moving it toward the minimum value associated with the Maxwell-Boltzmann distribution.

This was a result of astounding power. It implied not only that the Maxwell-Boltzmann distribution was uniquely the correct description of a collection of atoms at equilibrium but also that any other distribution would, because of atomic collisions, inevitably evolve toward the Maxwell-Boltzmann form. In fact, as Boltzmann was eager to believe, his quantity H was by all appearances exactly what he needed as a kinetic definition of the thing Clausius called entropy. All he had to do was put a minus sign in front of it. Then H reached a maximum in thermal equilibrium, was less than that for any other distribution, and starting from any other value naturally evolved toward equilibrium. This was how entropy behaved: whatever its value, it increased until it attained the maximum possible value, which corresponded to thermal equilibrium. H was precisely the kinetic definition of entropy, Boltzmann declared, and his H-theorem showed that the mysterious second law of thermodynamics, stating that entropy must always increase, was itself the consequence of the elementary principles of mechanics applied to the collisions of atoms. The H-theorem seemed to give a simple kinetic explanation for thermodynamics in its entirety. It seemed to be a proof, from first principles, of the inescapable fact that everything in the universe cools down and never spontaneously heats up.

But, as Maxwell was keenly aware, something about this result was fishy. In an 1869 letter to his friend and fellow-physicist Tait, he had come up with a whimsical character that came to be known as Maxwell's demon. Imagine two adjacent chambers filled with gas, one side hot and the other cold, and with an aperture connecting them. Normally, as atoms pass at random through the aperture, from one side to the other and back again, the gases will become mixed and their temperatures will become equal. Maxwell imagined, however, a tiny creature watching over the atoms passing back and forth and able to operate a shutter in the aperture. This demon's sole task was to open and close the shutter so as to let only the faster moving atoms into the chamber containing the hot gas and only the slower moving ones go the other way. The result of this monitoring would be to reverse the normal course of affairs: the gas in the hotter chamber would grow hotter still, and the gas on the other side would go colder. Heat would flow the wrong way.

Generations of undergraduates have been taught about Maxwell's demon, with often confusing consequences. The philosopher Karl Popper came to believe that Maxwell had somehow proved the laws of thermodynamics incorrect. There is, of course, no actual demon.

What Maxwell intended was a more subtle observation, which he made clear to Tait in the form of a catechism "Concerning Demons" of which item 3 read: "What was their chief end? To show that the second law has only a statistical certainty." The demon's actions contravened no law of physics. Maxwell's point, therefore, was that one could imagine atomic motions, no matter that they were somewhat fanciful, which resulted in heat flowing the wrong way.

Of course, the demon is a fictitious creature, a figment of Maxwell's imagination. But what the demon accomplishes on purpose can also happen by accident, without any demonic intervention. The probability might be exceedingly low, but it is not impossible that atoms would move purely by chance in such a way that heat flowed from cold to hot. Consequently, the second law of thermodynamics could not be an absolute law; there were circumstances in which it might not hold true.

This realization was the cause of Maxwell's skepticism over what Boltzmann claimed to have done. The H-theorem supposedly proved that any set of atoms, banging around and colliding at random, would move inexorably toward thermal equilibrium. But as Maxwell perceived, there must be physically permissible atomic motions that would correspond to heat moving the wrong way and, therefore, a system moving away from equilibrium. It might happen very rarely and only transiently, but it could nevertheless happen. Boltzmann's theorem, on the other hand, seemed to say that such a thing could not happen at all.

Maxwell therefore regarded the efforts of Boltzmann and others as basically futile. They were chasing after a mirage. Or, as he more extravagantly put it in his letter to Tait, "the German Icari flap their waxen wings in nephelococcygia amid those cloudy forms which the ignorance and finitude of human science have invested with the incommunicable attributes of the invisible Queen of Heaven." Nephelococcygia is Cloud Cuckoo Land, the foolishly idealistic city of birds, imagined by Aristophanes, that was supposed to exist midway between earth and heaven. In Maxwell's opinion, the German theorists did not understand that they were chasing after an unreliable fantasy.

Maxwell had, indeed, a peculiar sense of humor and an odd turn of phrase. A joking ironic manner was his habit. He had, like Boltzmann, lost a parent when young. Maxwell's mother had died, of intestinal or stomach cancer, when the boy was only seven years old. Young James's reaction was "Oh! I'm so glad. Now she'll have no more pain!" The boy was thereafter brought up by his father, a provincial lawyer, with the assistance of an aunt and an assortment of tutors. For the first 10 years of his life James Clerk Maxwell, an only child, grew up on the run-down estate of Glenlair, some 16 miles from Dumfries in the southwestern corner of Scotland, which his father had inherited. Here he explored the country and learned the stars, curious from an early age about everything around him. "Show me how it doos," he would ask at the age of three, his mother reported, and "what's the go o' that?" And if he didn't get what he considered an adequate answer he would insist "but what's the *particular* go of it?"

A streak of eccentricity ran in the Maxwell family. A grandfather reportedly saved himself from drowning in the Hooghly River in India by floating to shore on his bagpipes, which he then played to entertain the other members of his party and frighten away the tigers. Maxwell's father was similarly self-reliant and was endlessly fascinated by the ingenuity and invention displayed in all manner of novel industrial processes. He not only designed new buildings for his Glenlair estate, but had roomy square-toed shoes made up according to his own specification, and likewise had shirts cut to his personal preference. This was all very well when he and his young son were at Glenlair, but when young James was sent away to Edinburgh at the age of 10 to live with his aunt and continue his schooling at the Edinburgh Academy, his unusual garb and rustic accent caused a good deal of merriment among the more sophisticated city boys. They tore at his odd clothes, teased him for his outlandish and stuttering speech, and called him "Dafty." James showed a good deal of wit and resilience, and gradually earned the respect of his tormentors.

During his school years he wrote frequently to his father in zany letters filled with puns and misspellings, embellished with elaborate doodles, and containing secret messages in different colored inks. One letter begins: "My dear Mr Maxwell, I saw your son today, when he told me that you could not make out his riddles." And he took to signing himself anagrammatically as Jas. Alex. McMerkwell. His letters betray an affectionate and familiar manner, as well as some sharpness of mind. At the age of 11 he noted that "Ovid prophesies very well when the thing is over."

The boy showed early aptitude in science. When he was 14, some modest but original ideas of his in geometry were presented to the Edinburgh Royal Society. P. G. Tait, the physicist with whom Maxwell later corresponded, was a schoolfellow, and through his father, young Maxwell knew the Thomsons of Edinburgh, whose son William, later Lord Kelvin, was also to become a great scientist, engineer, and Victorian entrepreneur. He left the Edinburgh Academy in 1847 and moved on to the University of Edinburgh as an undergraduate. He was still only 16 years old.

Where Boltzmann was raised exclusively by a doting and perhaps overprotective mother, living with her until his marriage and only reluctantly separating even then, Maxwell moved between his father's residence at Glenlair, his aunt's house in Edinburgh, and the testing environment of the Edinburgh Academy–which, for all that Maxwell endured there from the other boys, gave him a solid education. He developed the habit of writing light verse and humorous doggerel, and commemorated the virtues of his first school:

> Let Pedants seek for scraps of Greek,
> Their lingo to Macadamize;
> Gie me the sense, without pretence,
> That comes o' Scots Academies.

(Macadam was the Scotsman who invented a durable road covering of stone chips and gravel embedded in a bituminous mix.)

Maxwell learned a good deal of self-reliance, bolstered by an ironic and occasionally waspish sense of humor. This attitude spilled over into his studies. As a student at the University of Edinburgh, he gave himself a summer project: "Kant's Kritik of Pure Reason in German, read with a determination to make it agree with Sir W. Hamilton." Sir William Hamilton was an Edinburgh professor of logic and metaphysics.

After three years at the University of Edinburgh, he traveled south to study at Cambridge. Here he was a misfit all over again, reading the lesson in Trinity College chapel in his wild Scots accent. He tended to speak, moreover, in a "spasmodic" manner, rushing out bursts of words then abruptly halting before delivering another burst.

But by this time his brilliance was becoming apparent, and Cambridge has traditionally admired eccentricity and idiosyncrasy when it is accompanied by the signs of genius. Maxwell, unable now to run about the Scottish moors for exercise, took to charging up and down the staircases of Trinity College in the small hours of the morning. His fellow undergraduates, recognizing the pattern, began to lay in wait behind their doors and fling shoes and hairbrushes at him as he passed. Maxwell made a number of friends with whom he kept close throughout his life.

He graduated from Cambridge in 1854 and stayed on to teach. After a couple of years he took a position in Aberdeen, then moved on to King's College London soon after. In 1865, at the age of 34, he withdrew from his academic position and began to spend more time at Glenlair, though he kept up his scientific work and maintained correspondence with his fellow physicists. Six years later, Henry Cavendish, the Duke of Devonshire and a talented physicist himself, endowed the University of Cambridge to build a laboratory for experimental physics, and Maxwell became the first director of what is still known around the world as the Cavendish Laboratory.

Maxwell's scientific endeavors showed great versatility. There were, moreover, great contrasts in the kind of theorizing he undertook. In his analysis of the rings of Saturn, he studied the mechanics of tiny particles in a gravitational field. In his electromagnetic work, he delved into pure field theory. In his work on the kinetic theory of gases, everything derived from mechanics; there was no field theory at all. This range of interests, compared to Boltzmann's more singular focus on gas theory, gave Maxwell a more agnostic view of theorizing in general. He found atomic theory fascinating, no doubt, and perceived its many virtues and possibilities. But at the same time he could see imminent difficulties.

BOLTZMANN, YOUNGER by 13 years, was introduced to Maxwell's work by Stefan, and throughout his life remained a great admirer of his Scottish counterpart. On one occasion, lecturing on Maxwell's electromagnetic theory, he borrowed a phrase from Goethe's Faust to ask rhetorically, "Was it a God who wrote these signs?" His respect was, however, only partially reciprocated. Maxwell and Boltzmann never met or even, apparently, corresponded. They might usefully have done so during the late 1870s, when objections to the H-theorem came up, but it perhaps seemed to Boltzmann that Maxwell had more or less withdrawn from the battle. Certainly, he didn't pursue the subject with Boltzmann's bulldog tenacity. Maxwell was guided by elegance and brevity and aimed to encapsulate in precise mathematics an idea or theory whose form he could already perceive. Boltzmann, by contrast, plowed on regardless, confident that because an answer must exist, it would be only a matter of time and effort for him to find it. He was never an introspective man, and in these circumstances that was an advantage; it never occurred to him to doubt whether he would succeed.

The differences in style between Maxwell and Boltzmann may have been in part consciously aesthetic, but were more likely the offspring of their individual psychologies. Both have their advantages and disadvantages. Boltzmann had doggedness coupled to passionate belief. Maxwell had a fierce sense of design or logic, which helped him find the powerful and beautiful simplicity of his electromagnetic theory. But, as Boltzmann's own work demonstrated, science is not always clean and precise, especially in its formative stages. Elegance is for the tailor and the shoemaker. Boltzmann's style enabled him to keep pushing ahead through the theoretical thorns and tangles of kinetic theory.

These differences also made Boltzmann, on his good days, a memorable lecturer and speaker, while Maxwell could find himself so anxious to convey the subtlety of each and every point that he lapsed into inarticulate stumbling and then silence. Though he complained with increasing frequency, as he got older, of the tedium of lecturing to dull and indifferent undergraduates, Boltzmann was capable of expounding the subjects he cared about with a passion unconstrained by doubt or hesitation.

Maxwell, by contrast, was not a man of overt exuberance. There was in him almost something of the dilettante, a lighter touch that enabled him to hop from one subject to another, as he did throughout his scientific career, but that made him easily distractible when attempting to teach. He had difficulty keeping his mind on a single track and, as one contemporary put it, he had, as a result, "his full share of misfortunes at the blackboard." At one point both Maxwell and Tait applied for the same position at the University of Edinburgh, and although Maxwell was acknowledged to be the greater scientist, Tait got the job, it seems, because he could teach.

If, in their spoken expositions, Boltzmann was forceful and Maxwell hesitant, their writings show their characters in a different light. Maxwell thought and analyzed a great deal before he committed anything to paper, trying to work his way through every possible byway in advance. What he wrote was therefore clear and complete, carefully and logically guiding the reader to an inescapable conclusion. Boltzmann, characteristically, wrote as he spoke, forging his way ahead without worrying that every possible byway had been inspected and every possible objection assessed and discarded. His bulldozer style made his writings often dense and difficult and–disturbing to his readers and often to himself–not always consistent from one exposition to the next.

Maxwell had the essence of the ironic perspective: the capacity to stand away from his own work and see it as others might see it. He could write persuasively because he could anticipate objections from those with another point of view, and answer them before the detractor had clearly articulated the problem. Boltzmann, throughout his life, in his personal dealings and in his science, was deaf to other sensibilities. "I think I have established my priority," he would write, not understanding that the reader was already there.

In yet another letter to Tait, Maxwell expressed his opinion of the differences between them: "By the study of Boltzmann I have been unable to understand him. He could not understand me on account of my shortness, and his length was and is an equal stumbling block to me. Hence I am very much inclined to join the glorious company of supplanters and to put the whole business in about six lines." He was admittedly reading Boltzmann in German, but then Maxwell had read Kant in German as a teenager, so he had the necessary fortitude.

This was written in 1873, the year after Boltzmann published his H-theorem, which Maxwell found impossible to understand because of his own ideas concerning what he called the demon. He could not see how Boltzmann had been able to derive an equation demanding a one-way trend in the behavior of atomic motions, when clearly there must be sets of atomic motions that behaved differently. And so Maxwell concluded that somewhere in the dense reasoning of his 1872 paper Boltzmann must have gone wrong.

Boltzmann did not immediately perceive the acuteness of Maxwell's criticism. A detailed account of the demon appeared in Maxwell's *Theory of Heat,* which was published in 1871 and translated into German in 1877. But by that time the same issue had been raised in a somewhat different guise, and in circumstances such that Boltzmann could hardly evade the question. The source of the objection this time was his friend and colleague Josef Loschmidt.

What Loschmidt articulated, in 1876, became known as the reversibility problem. It hinges on the fact that the laws of mechanics governing atomic motions and collisions are, as physicists like to say, time-reversible; that is, any set of motions and collisions obeying Newton's laws can be run backward, as if on a videotape, and they will still obey Newton's laws. This, as Loschmidt explained, leads to a problem with the H-theorem: any set of atomic motions that causes H to decrease has a time-reversed counterpart that must cause H to increase. How then can Boltzmann's theorem dictate that H must always decrease? Loschmidt, a sympathizer of atomic theory, did not intend his observation as a disproof of Boltzmann's work in particular or kinetic theory in general. But undoubtedly he had located a problem that demanded an answer.

Loschmidt's reversibility objection was essentially what Maxwell had in mind with his somewhat cryptic invocation of the demon. But Maxwell was perhaps too cute for his own good. Thomson had published something very close to Loschmidt's argument a couple of years earlier but seemed to conclude that in the absence of the hypothetical demon, these oddities would not arise.

Faced now with Loschmidt's specific objection, presented moreover to the Viennese Academy of Sciences, Boltzmann had to come up with an equally specific response. His first answer was simple. He agreed, as he must, that for some atomic distributions the value of H, and therefore the entropy, must go in the wrong direction. But he asserted that such cases would demand an extraordinary degree of order–a sort of conspiracy–among the atoms. As a matter of probability, and because of the enormous number of disorderly arrangements of atoms compared to "special" arrangements, H would almost always do what the H-theorem said it would do.

All kinds of traps and implications lay concealed in this answer. For one thing, Boltzmann had originally said that the H-theorem was exact, that atomic collisions would always lead to an increase of entropy. Now he was saying that in certain cases, rare, unlikely, but physically legitimate, this would not be so. But in that case was the H-theorem a true theorem, a theorem with limited validity, a useful approximation, or what exactly? And if the H-theorem did not always hold true, what was the extent of its validity, and what precisely was the nature of atomic distributions for which it wasn't true? Boltzmann had derived the theorem in a seemingly general way, using basic elements of Newtonian mechanics and some broad, seemingly plausible arguments about the behavior of atoms. Was there some hidden assumption that was not quite true, or not always true, so that the H-theorem would not invariably follow?

Furthermore, if Boltzmann was now saying that in some odd cases entropy could decrease instead of increasing, was the implication that in nature, in reality, there could be occasional instances of systems behaving counter to the recently minted second law of thermodynamics, or was he implying that the atomic distributions that gave the "wrong" behavior were, for some as-yet-unspecified reason, physically disallowed? Critics seized on the suggestion that kinetic theory seemed to imply that the laws of thermodynamics were not true laws after all, but only approximate laws, true "almost always." If this was so, it seemed an unhappy development. There had never been any implication that Newton's laws of mechanics were true only most of the time, or that the refraction of light by lenses almost always went according to plan. What use was there–indeed what meaning–in a supposed law of physics that on closer examination turned out not to be quite a law after all?

Critics of atomic thinking, who were beginning to rally against the seeming triumphs of kinetic theory, now believed they had found a crucial flaw. They held the laws of thermodynamics to be absolute and inviolable, as true physical laws must surely be. Kinetic theory stumbled on this point. Taken literally, it implied that the second law of thermodynamics was inexact and therefore not really a law at all. Or if, as Boltzmann sometimes seemed to hint, kinetic theory must be amended or augmented somehow so as to disallow violations of the laws of thermodynamics, then its pretensions to be a complete explanation of thermodynamics based only on mechanics were demolished. Either way, atomic theory looked shaky.

Maxwell, the first man to perceive the probabilistic nature of the second law, doubted Boltzmann's theorem because it seemed to offer absolute certainty where there could be none. Thomson, on the other hand, who had originally praised Boltzmann's work because of the very certainty it promised, now began to have doubts because probability was creeping into the picture.

In Germany and Austria the prevailing opinion was with Thomson. The laws of thermodynamics must be absolute, so kinetic theory must be wrong. And beginning to be influential was the voice of Ernst Mach, still in Prague but making a few inroads with books of historical and philosophical commentary on physics. As a student in Vienna, just a couple of years ahead of Boltzmann, Mach had been swayed by atomism and counted himself for a time a believer in atomic theory. But in Prague he was finding his own voice and was beginning to evolve a philosophy of science according to which observations and data were primary and theorizing was intrinsically suspicious. The goal of science, Mach implied, was to provide logical and rational relationships between facts and phenomena that could be directly observed; the more one invoked the existence of entities whose existence was not immediately apparent, the more one was going astray. Theorizing, in Mach's view, was a necessary evil at best, and frequently an unnecessary one.

In atomism and kinetic theory, Mach found a natural target. It demanded a belief in unseen and quite possibly unseeable objects, yet its results, which merely confirmed what the laws of thermodynamics already said, were supposed to lend credence to the assumptions on which it was based. Apart from the circular nature of this reasoning, it ran counter to what Mach had decided was the essence of scientific explanation: to find laws, as simple as possible, linking observable phenomena. Classical thermodynamics passed the test; it posited fundamental relationships between the overt properties of gases–their pressure, volume, temperature, and so on. Kinetic theory, on the other hand, sought to replace these perfectly acceptable and straightforward laws with new and mysterious explanations based on unprovable assumptions about the existence and properties of atoms. How was this an advance?

The discovery of flaws, paradoxes even, in the kinetic theory represented not just problems for the theory itself–problems that Boltzmann at least thought he knew how to deal with–but deeper flaws in the very structure and essence of the theory, so far as Mach was concerned. To tinker with the theory so as to bring it into line with established thermodynamics laws was, Mach concluded, an admission of failure. Proponents of kinetic theory had originally proclaimed that armed with nothing but the laws of mechanics, they could explain the properties of gases. Now they discovered that perhaps they could not, so they began to modify their already unfounded theoretical assumptions.

To Mach, the conclusion was simple. Atomic theory had failed in what it professed it could do and must therefore be wrong. His distaste for theorizing was vindicated. His insistence on sticking to simple laws linking observable data was shown to be reliable. The kinetic theory of heat, in the view of Mach and those who were beginning to rally around him, had had its day.

CHAPTER 5

"You Will Not Fit In"

The Daunting Prussians

ALTHOUGH BOLTZMANN EMPLOYED novel techniques of mathematical statistics in establishing his H-theorem, he did not immediately appreciate that the statistics went deeper, into the physics itself. To Maxwell goes the credit for first understanding that the second law of thermodynamics is inherently a matter of probability: it is only unlikely, but not impossible, that heat should flow from a cold body to a hotter one. Spurred on by his friend Loschmidt's objection to the H-theorem, Boltzmann quickly made up the deficiencies in his understanding, and it was he, not Maxwell, who succeeded in quantifying the improbability of heat flowing the wrong way—of specifying, in other words, just how unlikely it was that the second law would be violated.

In tackling this problem, Boltzmann navigated through the intricacies of kinetic theory to an entirely new perspective on the subject. Mechanics was still the key ingredient. Atoms moved according to Newton's rules, and therein lay a true understanding of the properties of gases. But analyzing the mechanics of atoms was a formidable undertaking, and Boltzmann saw that he could look at the whole problem from a different angle, by bringing probability and statistics to the fore. He had already begun to think of the state of a gas—the set of atomic motions represented by the velocity distribution at any one moment–as an important theoretical conception in its own right. Now he began to formulate a new mathematical treatment of these states, relying on the mathematics of probability more than on the familiar differential equations of mechanics.

At any moment, the atoms in a volume of gas must have a certain set of energies; the next moment, the energies of most of the atoms would have changed, because of collisions, and the whole collection would be characterized by a new set of energies. Each set of energies constitutes an individual state of the gas as a whole, and the result of incessant atomic collisions is that the gas skips constantly from one such state to another. Boltzmann set about constructing a calculus of probabilities in which these states of a gas form the basic elements.

An immediate difficulty was that the amount of energy any atom possesses at any moment is an infinitesimally variable quantity, definable exactly only if one is willing to write the number down to an infinite number of decimal places. Consequently, there must be an infinite number of possible distributions of energy among even a finite set of atoms. To make the problem manageable, Boltzmann hit on the idea of dividing the range of possible atomic energies into a set of pigeonholes of some finite size–as if, for example, one decided to specify the energies only to three decimal places, and classify as equivalent all energies that differed only in the fourth decimal place or beyond. Specifying the state of a gas, in this new system, means listing the number of atoms in each pigeonhole. As the state of the gas changes, atoms hop out of one pigeonhole and into another, but always there will be the same number of atoms in total, scattered in some way among all the available pigeonholes.

This picture offered some new insights. From a specific arrangement of atoms in pigeonholes, for example, pick two atoms at random and swap their places. Doing so creates a different state, but since the number of atoms in each pigeonhole is unchanged, it's a state with the same physical characteristics as the first one. This clearly illustrates the important fact that a large number of possible distributions of atoms can correspond to a volume of gas with the same physical properties.

Pursuing this idea, Boltzmann saw how he might obtain from it quantitative results, not mere pictorial assistance. He imagined taking a certain number of atoms and scattering them at random into pigeonholes, with the only additional stipulation that the total amount of energy carried by all the atoms together was some fixed quantity. This was equivalent to stipulating a quantity of gas with some fixed total amount of heat.

He then set himself to analyzing the likelihood of all the possible distributions of atoms in pigeonholes. It is unlikely, one would think, that all the atoms would end up in one pigeonhole, or in just a few, and much more likely that they would be fairly evenly scattered among all the available slots. But why, precisely, is it unlikely that all the atoms should occupy one pigeonhole? Because, Boltzmann realized, there's only one way of doing it: every single atom has to go into the same spot; there's no other choice. By contrast, when the atoms are scattered among a number of different pigeonholes, the resulting pattern is more likely because there are more ways of achieving the same result: it doesn't matter whether atom A is in slot 1 and atom B is in slot 2, or vice versa, as long as the same overall distribution of atoms stays the same.

This crucial insight led Boltzmann to perhaps the crowning achievement of his career, and to a result that stands today as a monument to his scientific insight. In a publication that appeared in 1877, he showed how to measure the probability of atomic distributions by calculating the number of equivalent ways they could be constructed. This led to the first important finding: the most likely distribution was yet again the one dictated by the Maxwell-Boltzmann formula. Thermal equilibrium, in Boltzmann's new analysis, emerged as simply the most probable way of sharing a fixed amount of energy among a fixed number of atoms.

But there was more. With the same method, Boltzmann could calculate the likelihood of any distribution of atoms in pigeonholes, no matter how peculiar. The closer the pattern was to the optimal, equilibrium one, the more likely it was; the farther from the optimum, the less likely. Here again was a connection with entropy, which measures how close any distribution is to equilibrium. Boltzmann formulated what was to become, for physicists anyway, a famous and simple equation, which declared that the entropy of any distribution of atoms is proportional to the logarithm of the number of equivalent ways that distribution can be constructed. In accordance with the principle that a book intended for a general audience is allowed one equation, Boltzmann's formula deserves to be stated here in its standard modern form:

$$S = k \log W$$

S is entropy, W stands for the number of possible ways of making a given distribution of atoms in pigeonholes, and log means logarithm. (The log is the inverse of an exponential function, increasing as W increases but more slowly as W gets larger.) In 1877, Boltzmann did not have an accurate determination of the number of atoms in a standard volume of gas, and therefore stated his result simply as a proportionality between S and log W. The quantity k, now known as Boltzmann's constant, was defined later.

This straightforward formula, the gem unearthed by Boltzmann in 1877, offers a completely new way of calculating entropy and a new way of understanding it. It corresponds closely to the quantity H that he had derived five years earlier, but whereas H came from a careful consideration of the motion and dynamics of incessantly colliding atoms, S emerges from what seems almost a childish game of throwing atoms in pigeonholes. The more ways there are of creating a certain distribution of atoms, the greater the entropy of that distribution. Yet both S and H denote essentially the same thing, and both, Boltzmann demonstrated, are equivalent to the thermodynamic entropy defined by Clausius in 1865 from the physical properties of an actual gas.

Perhaps the most remarkable thing about Boltzmann's 1877 result is that it seems almost devoid of physics. Implicit in the calculation of H is the notion of atoms banging and crashing ceaselessly into each other until they reach a stable distribution: thermal equilibrium. The increase of entropy is then directly the result of mechanics. But in the definition of S, any such mechanical picture of atomic motions has apparently vanished. Boltzmann is able to formulate entropy by thinking purely in terms of possible states or arrangements of atoms, without any regard to where these arrangements come from or what they develop into.

Still, the physics has to be in there somewhere, and it is. The endless collisions between atoms in a gas cause the state of the gas–the distribution of atoms in pigeonholes–to be endlessly in flux. Boltzmann's argument is that as a gas settles toward equilibrium, its constantly changing state explores all the possibilities open to it and, simply as a matter of probability, ends up spending most of its time in the most likely arrangement. The implicit and rather subtle assumption is that the gas indeed obeys a sort of equal opportunity rule, rattling around the possible states in an evenhanded manner, and having no special or preferred state. It is this assumption, which Boltzmann certainly recognized, that permits the use of simple probability calculations to determine in what class of states a volume of gas will spend most of its time.

This crucial assumption, that all the states of a gas are equally likely to be visited as atoms move and collide, swallows up all the tricky but essential questions of mechanics. The way the atoms change from one overall state to another is fundamentally a question of mechanics–the very question Boltzmann tackled in producing the H-theorem. Now he was putting all those complications aside and simply assuming that the atomic states would visit unpreferentially all the states open to them.

It seemed a plausible enough idea, but this equal-opportunity rule was hard to state precisely, let alone explore theoretically. Strictly speaking, it could not be exactly true. Defining every individual atom's state of motion to the last decimal place would demand infinite precision and would imply an infinity of possible states. Clearly, in a finite amount of time, atoms could not visit every such state. But Boltzmann's analysis involved putting atoms into the appropriate pigeonhole according to their approximate energy, and by the same token the practical necessity was that atomic states should get close enough to every possible state in the course of a reasonable period of time. But how close is close enough, and how long a time is reasonable?

These were murky questions. Boltzmann's 1877 paper was, therefore, profound, far-reaching, and baffling all at once. Profound because it offered a definition of entropy based on a simple argument about chance and likelihood; far-reaching because that definition allowed entropy to be calculated for any system where there was some measure of order or disorder, such as, in modern times, digital information and communication; and baffling because it rested on an assumption that seemed difficult to state, let alone prove.

There was also, necessarily, a version of Loschmidt's argument hidden in the new picture. Entropy increased, Boltzmann was now saying, because any system tended to move from less likely to more likely configurations as time passed. That was what "likely" meant, after all. But this is a tendency, not an iron rule. Just as Loschmidt had observed that atoms could at least on occasion move so as to make Boltzmann's H go the wrong way, so it was possible that, from time to time, a system might happen to move briefly into a less likely state, thus making entropy temporarily decrease rather than increase. But now it became possible to put a number on the probabilities involved. For any given state, Boltzmann's methods specified how many states were more likely and how many less likely. If there were a million times more of the former than the latter, then there would be only a one in a million chance that the system would move in the "anti-entropic" direction. In practice, calculations of this sort were hardly feasible. Still, Boltzmann could feel once again that he had solved the essential riddle and shown precisely what it meant to talk of the improbability of heat flowing the wrong way.

Boltzmann published his statistical formula for entropy the year after he was married and had moved back to Graz. After a pause in his work on kinetic theory, he was back in full flow. That same year, 1877, he published a total of five scientific papers, four of which concerned the theory of atoms and molecules (the other was a short mathematical work). Although his work moved on seemingly without letup, he found that the circumstances of his life were not as simple as they had been. He was no longer the peripatetic scholar, devoted to nothing besides his work. In a jokey but nonetheless revealing letter to Toepler written after he had been in Graz just over a year, he remarked on the change. Whereas before he had simply needed a room to eat and sleep, now Henriette required something rather more elegant. "I fear that my earlier principal was correct," he went on, "and that one's achievements are in no way proportional to the size and splendor of one's surroundings. I had expected that marriage might make a person somewhat lazier, but to what extent this would turn out to be true I would not have believed."

Six months later, writing again to Toepler, Boltzmann allowed that things were going pretty well, but lamented that he was so far from the center of scientific activity and that there was consequently a lack of intellectual stimulation for him. Over the years he lived in Graz with his wife and growing family his letters to Toepler carried a stream of minor complaints. His health bothered him occasionally–a letter in 1879 contains the first mention of asthma, which was to become a recurrent plague for him–and his failing eyesight made it necessary for him to apply for a little extra money so he could hire a scientific assistant to help him with arduous calculations.

He complained more than once of his sense of isolation, and that he had become estranged from his former colleagues in Vienna, especially Loschmidt. But this isolation was in large part his own doing. He failed to keep up any correspondence with Stefan or Loschmidt, and having made the trip between Graz and Vienna countless times during his courtship, he now ceased abruptly. After his honeymoon in Switzerland, six years went by before he visited Vienna again.

At times he also complained of his administrative and especially his teaching duties. He had an assistant in the person of Albert von Ettingshausen, a nephew of Andreas von Ettingshausen, who had been Stefan's predecessor as director of the physics institute in Vienna. The younger von Ettingshausen helped with the teaching of experimental physics in particular, and took care of a good deal of the day-to-day bureaucratic demands that running an institute and teaching students necessarily involved. But still a considerable teaching load fell on Boltzmann's shoulders. The education ministry had noted that his appointment was a two-for-one deal, in that the young physicist was a competent experimenter as well as a distinguished theorist. But having helped him land the job in Graz, these talents made the job itself onerous. For many years he taught experimental physics and laboratory classes in addition to lecturing across the range of mathematical physics.

And he wanted to do a good job. He was obliged to teach introductory classes for the numerous students of medicine and pharmacy who came to Graz, and he earnestly wanted even these young men, who were clearly not destined for a life in research, to understand what he was telling them. "He did not rest," a colleague later observed, "until he had convinced himself that the course was being followed by everyone in the audience." This diligence took its toll. As the young chemists Nernst and Arrhenius discovered, Boltzmann was so occupied with these elementary lectures that he found it hard to make time for students who were interested in higher matters. Still, as Nernst and Arrhenius also found, Boltzmann would spend hours with them when he realized their fascination with science was as great as his, and so he used up even more of his time.

Despite his rueful admission to Toepler that marriage had made him lazy, he continued to publish at an increasing rate during his first years in Graz. After coming back to kinetic theory with his enunciation of the statistical formula for entropy, he set the subject largely aside again, devoting his energies to numerous investigations of electrical and magnetic phenomena. But then he came back to kinetic theory with a series of papers exploring diffusion and viscosity in gases in molecular terms, and followed that with what is now called the Stefan-Boltzmann law, his thermodynamic explanation of the energy and pressure of electromagnetic radiation. Perhaps he was, as some of his colleagues remarked, a shy and reclusive man, but the sheer volume of work he undertook in these years, along with the demands of his young family, would have caused many a man to all but disappear from the social scene.

THROUGH THE 1880s, the deeper implications of Boltzmann's two great achievements–the H-theorem and the statistical definition of entropy–went largely unexplored. Especially in the German world, the prevailing opinion was that the laws of thermodynamics must be absolute, so that any talk of probability was fundamentally mistaken. Critics of kinetic theory simply ignored Boltzmann's work, and for the time being Boltzmann himself had neither the time nor the energy to proselytize.

The one man who might have been his ally, or spurred him on, was himself ambivalent. In 1878, Maxwell published an analysis of what he called "Boltzmann's theorem"–the only time he mentioned his Austrian counterpart in the title of one of his own works. But what concerned Maxwell here was not the H-theorem but Boltzmann's much older argument, of 1868, in which he developed Maxwell's formula for the distribution of atomic velocities into the more general Maxwell-Boltzmann distribution, which applied for atoms possessing energy of any sort, not just motion.

In this generalization, the ever-astute Maxwell had realized, lurked essentially the same "equal-opportunity" rule that Boltzmann had more explicitly assumed in his proof of the statistical entropy formula. In other words, as atoms moved and interacted and exchanged energy, it was important that they exchange all kinds of energy equally, in the long run, so that the numbers of atoms with particular energies didn't depend on what form that energy happened to take. This was, and remains even today, a difficult and subtle question. In 1878, Maxwell ruminated on it further and clarified some of the issues involved, but he fell short of proving that the rule was inevitably true.

This somewhat inconclusive analysis was Maxwell's last pronouncement on kinetic theory. His health had begun to fail in 1877. His digestion troubled him, but for a year or more he did little except dose himself with bicarbonate of soda and keep quiet about the pain. There was in any case little his doctors could have done for him. As a boy he had seen his mother die painfully of intestinal cancer, and now the same fate was befalling him. He continued to work intermittently, when he could, during the next two years, and he continued to write wry and humorous letters to his friends and colleagues. He died in November 1879, only 48 years old.

Maxwell had once written to Tait that "by the study of Boltzmann I have been unable to understand him," and it seems that he was never able to muster the energy needed to work through Boltzmann's longer arguments in order to discover how their puzzling conclusions came about. There are signs in his last paper that he might finally have been getting around to the task, and the premature end of his attempt was a misfortune for physics in general and for Boltzmann in particular. For the rest of his life, criticism of his statistical methods and confusion over the meaning of his theorems pursued Boltzmann, and he, characteristically, had endless difficulties answering his detractors in a way that they would find persuasive. Maxwell's rigor and clarity might have helped.

Over the next decade, other deaths were to afflict and influence Boltzmann's life. In January 1885, Boltzmann's mother died just four days after her 75th birthday. Boltzmann was then a couple of weeks short of turning 41 years old.

Although he had by this time been married almost 10 years and had four children of his own, his mother's death unmoored him. He was depressed and anxious; his steady and reliable rate of publication came to an abrupt halt. That year he published only one piece of scientific work, and no letters written by him are known.

His mother had from the beginning guided his education and his career and had thrown her own life into his, devoting the family's resources to the brilliant son. She took no interest in science, beyond pride that her son was associating with the great names of Europe, but some part of Boltzmann's continuing achievement was still in her name, in honor of all the effort on her part that lay behind his own success. Now she was gone. Although Boltzmann relied on his wife in many respects, and deferred to her in domestic matters, he seems never to have confided any of his intellectual and scientific struggles in her. His children were young. He had no truly close academic relationships, and Graz, agreeable though it was, seemed a long way from Vienna, let alone Cambridge or Berlin. His life's work, kinetic theory, had produced some early triumphs but now seemed to have reached an impasse, fraught with confusing questions to which Boltzmann had not yet worked out thoroughly convincing answers. In the middle of his career and his life, Boltzmann was respected but frustrated, accomplished but uncertain.

October 1897 saw the death of another eminent physicist, Gustav Kirchhoff. The two had had little contact since their first meeting in Heidelberg some 20 years earlier, but Boltzmann maintained a fond recollection of that encounter and admired Kirchhoff as a physicist. Kirchhoff had in the meantime ascended to one of the highest peaks of German physics, becoming a professor in Berlin, and the immediate impact of his death on Boltzmann was that an important position now became vacant in one of the most important universities in the German-speaking world.

The search for a successor to Kirchhoff in Berlin naturally embraced Boltzmann. He was among a handful of theoretical physicists of worldwide reputation. If few other physicists had read and understood his publications on kinetic theory, and if his ideas still caused controversy, his work was nonetheless regarded as the work of a master. An internal document recommending him for the Berlin position described Boltzmann as "a highly astute, outstanding mathematician who has succeeded in solving some of the hardest and most abstract problems" in kinetic theory. And his other researches, on electricity and magnetism and on mathematical physics in general, constituted a solid if not spectacular scientific achievement in their own right. He had, furthermore, visited Berlin, and knew Helmholtz, insofar as anyone could be said to have known the frosty "Reichschancellor" of German physics. They had continued a sporadic correspondence during Boltzmann's years in Graz.

Toward the end of the 1880s, Boltzmann had also begun to accumulate a variety of reasons to think about a move. For the academic year beginning in 1887, Boltzmann served as rector (that is, president) of the University of Graz. He was in charge, at least nominally, of its entire administration. Boltzmann had little interest or aptitude in such matters, but in the course of his tenure at Graz increasingly important offices came to settle on his shoulders whether he liked it or not. As rector, Boltzmann had to deal with an episode of student unrest, arising out of growing nationalist tensions within the Habsburg realm. Groups of German students began to express their resentment of all the non-German elements within Austria-Hungary, not just the Hungarians but also the Czechs, the Slovenes, the Poles, the Serbs. These student groups held as their ideal a "Greater Germany" that would encompass all the German peoples of Europe, including those in Austria, and they looked to the kaiser in Berlin rather than the emperor in Vienna as their natural leader. In November 1887, a drunken riot broke out at the university in which busts of Franz-Josef and his empress disappeared.

All professors in Austrian universities served at the discretion of the ministry of education and, ultimately, of the emperor himself. As Boltzmann strove to settle the uneasy atmosphere at Graz, feeling obliged to punish some but not wanting to incite more protest or violence, his actions were being observed from Vienna. Boltzmann was later reported to have said, in connection with nationalist protests among the students in Graz and elsewhere, that he had once noticed of the pigs on a farm he had visited when young that the tails of some curled to the left and others to the right, but that he did not know "if the pigs with the left-curled tails grouped together against those with the right-curled tails or not." Sage though such remarks might be, they can have been of little value in calming the antagonisms between exuberant and sometimes fanatical groups of students.

With, in all likelihood, the considerable assistance of seasoned administrators, the troubles at Graz were laid to rest at least for the time being. But these events diminished any view of Graz as a place of calm repose. After a generally happy decade there, Boltzmann's quietude was beginning to fray.

Boltzmann, approached for the Berlin position, visited the Prussian capital and was treated with great respect and deference. Even for someone of Boltzmann's acknowledged prowess, the possibility of a move to Berlin, the powerhouse capital of Prussia, must have been intoxicating. In January 1888, in Berlin, Boltzmann signed a letter of intent to take up an appointment there the following fall. Matters of salary, duties, accommodations, and so on, had been agreed to. With this letter in hand, the Berlin authorities set about obtaining formal clearance for the appointment, which needed, ultimately, the approval of the kaiser, just as similar appointments in Vienna and Austria needed the signature of Franz-Josef. For so prestigious a position, the kaiser no doubt took more than a routine interest in who was being sought out.

Back in Graz, word of Boltzmann's visit to Berlin, and the reason for it, made it into the local newspapers, although the fact that he had actually put his name to a letter taking the job remained a secret. A couple of months later, in March, the kaiser appended his name to a formal offer. In the meantime Boltzmann's long-term assistant, Albert von Ettingshausen, decided that if Boltzmann was about to leave he would be better off finding another safe harbor, and in February announced that he was leaving the University of Graz to become a lecturer in physics at the Technical High School. Heinrich Streintz, another physics lecturer at the university, began to approach the administration about obtaining more funds for his own work. Boltzmann's imminent departure led to all kinds of jockeying as other physicists, for so long in his shadow, made their moves to emerge into the light.

But all was not what it seemed. When word reached the authorities in Vienna that Boltzmann was being courted by Berlin, the question of national pride erupted. Officials of the Habsburg court can have had little idea what the work of this physicist meant, but they could see for themselves the forbidding pages he had published in the *Proceedings of the Viennese Academy of Sciences,* and they knew that his work was esteemed highly by scientists in Cambridge, Heidelberg, and Göttingen, as well as Berlin. The very fact that Boltzmann was being sought by the Prussians must mean that he was a great man, and so it became important to keep him in Austria, whether in Graz or elsewhere.

Boltzmann was quickly informed that the education minister "would regret it most keenly if your honor's excellent services were to be lost to the University of Graz and to the Fatherland, and would place great value on your honor's remaining in his present position." Boltzmann offered his assurance to a visiting ministry official that he had not yet made any firm decision, but it was made clear that a written statement to that effect would be much appreciated.

Now he was in a fix. He had signed a letter in Berlin saying he would start his new position there that fall. He had told Vienna that he had made no such commitment, and now he was being asked to say so in a letter. Given the level of interest in Boltzmann and his choice of appointment, he was in effect telling the kaiser one thing and the emperor another.

There is, incidentally, a small linguistic sign hinting at the delicacy of Boltzmann's position. In English, a university makes an offer of an appointment; in German, the university sends out a "call" (*Ruf*) to someone it wishes to hire. In the 19th century, that call would come from the royal court of Berlin or Vienna, in the monarch's name. Declining an offer is not quite the same thing as refusing an imperial call.

The charitable explanation for Boltzmann's frankly deceptive behavior is that he had signed the letter of intent in Berlin knowing that the kaiser's approval would take some time, and so figured he was merely committing himself to a possibility, not a certainty. Then, put on the spot in Graz by the emissary from Vienna, he avoided telling the truth out of a mixture of nerves and embarrassment. Somehow, he must have reckoned, he would work things out.

Some inner hesitation prevented Boltzmann from following through on his acceptance of the Berlin offer. One anecdote, recounted years after Boltzmann's death, attributes a fateful phrase to Helmholtz's wife. On his visit to Berlin, Boltzmann was at dinner one evening with Professor and Frau von Helmholtz and had picked up the wrong piece of cutlery. The professor's wife turned to Boltzmann and said, "Herr Boltzmann, in Berlin you will not fit in." Boltzmann himself, many years before, had remarked that in physics he respected Helmholtz more than any other, but that personally he found him difficult and remote. Now the great man's wife was reinforcing Boltzmann's unease.

Under pressure from the Austrian ministry, Boltzmann abruptly wrote to the Berlin authorities trying to beg off his impending appointment. His eyesight, always bad, was getting worse, and he regretted to have to say that he could not be sure of taking on the teaching load required of him. Berlin, to Boltzmann's dismay, was wholly sympathetic. They had excellent doctors, and they would provide assistance to Boltzmann in whatever way would help.

Vienna, in the meantime, offered an increased salary to keep Boltzmann in Graz. His first ploy having failed, Boltzmann wrote to Berlin in early June of 1888 to say he was seeking his release from the Austrian ministry (he needed the emperor's signature to leave an appointment in Austria as well as to assume one). But on June 24, the desperate physicist wrote two strange letters to the kaiser's officials. One reiterated the difficulties connected with his poor eyesight but went on to express his own lack of confidence about his abilities to teach mathematical physics in so distinguished a venue. Bizarrely, wrote the foremost practitioner of theoretical physics in the German-speaking world, "by accepting my new appointment in

Berlin I would be embarking on a wholly new field, that of mathematical physics, whereas here in Graz I have taught mainly experimental physics, and that indeed to students of medicine and pharmacy." With this letter he included two medical reports, one from his eye doctor, who described the parlous state of Boltzmann's vision and its poor prognosis, the other from the well-known psychiatrist Richard Krafft-Ebing, who was then a professor at Graz. Krafft-Ebing testified to Boltzmann's weak psychological state, citing neurasthenia and general nervousness. Boltzmann begged to be released from his promise to take up the Berlin position.

The other letter was a more informal and personal note, written to one of the Berlin officials he had been corresponding with. It was short and to the point. He admitted that his nervous agitation had increased greatly; though he had earlier been overjoyed at the prospect of the Berlin position, he had now come to the conclusion he would be an unworthy successor to Kirchhoff. And he apologized for all the trouble his indecision had created.

And then he changed his mind again, and then again. On June 27, he sent an urgent telegram to Berlin asking that his letters of June 24 be returned to him unopened; the very next day he sent another telegram countermanding his message of the day before, now asking for that message to be ignored and the letters opened after all. At some point, the letters were indeed opened and the Prussian authorities, possessed now of a set of frantic and contradictory communications, found a way to consult Boltzmann's wife via a family friend, one Franz Schulze, who had been professor of zoology in Graz and was now in Berlin. Had Boltzmann approached Vienna about being released from his position in Graz, they inquired, or was it possible that he could still remain there? She replied that on account of his nerves, already worn down by his duties as rector and now aggravated further, her husband had in fact still made no attempt to obtain a formal release. She hoped that over the summer he would have a chance to recover his strength. But her letter left unclear what might be Boltzmann's best course. "Since I have seen for myself how dreadfully hard it hit him to give up the Berlin position," she wrote, "I fear, unfortunately, that he will feel unhappy for the rest of his life to have forfeited forever a position that would so well suit his preferences. There is hardly anything that can be done!"

Digesting this information, Schulze told the Berlin officials that his "poor friend would only attain peace when the agony of deciding is behind him" and recommended releasing Boltzmann from his obligation. On July 9, the kaiser officially did just that. After six months stricken by nerves and confusion, Boltzmann was now free.

But he could not easily let go. Boltzmann's failed attempt on Berlin was the first time in his life he had not succeeded in achieving something he thought he deserved, the first time he confessed to any doubt over his own abilities. In his theoretical work he had successfully plowed through difficult problems, confident that he would find his way to the solution, and worrying about logic and consistency afterward. It never occurred to him that a problem could defeat him, that he might not succeed. He had adopted something of the same manner when, appointed a mathematics professor in Vienna, he had taught his own selection of theoretical physics, getting permission as and when he needed.

Muddling through–*fortwürsteln*–was an acknowledged Austrian characteristic, if not quite a virtue. The Prussians were not muddlers. While Franz-Josef struggled endlessly to keep Austria-Hungary in one piece in the face of constant bickering among all its disparate peoples, Prussia, guided by Bismarck, set about consolidating its position as a power in Europe through strategic alliances and stealthy expansion. In neither case was any powerful ideology at work. Franz-Josef muddled along, trying to placate each part of his squabbling empire in piecemeal fashion, as and when he could. Bismarck equally was capable of forming and reforming his alliances, shaping his tactics as circumstances demanded, but he and the kaiser had a sense of purpose, a national goal. They were out to achieve greatness while their Austrian counterparts were fighting to prevent disintegration. Berlin was rising and Vienna was falling.

Despite their militaristic reputation, the Prussians were not necessarily hard people to deal with. Michael Pupin, the American physicist who had at first found Helmholtz and his colleagues reserved and unapproachable, warmed up to the people of Berlin after a few months there. They were honest and straightforward, he came to think, and their stiffness derived from a sense of dignity. In their dealings with Boltzmann–who must have seemed a strange and slippery creature to them–they showed a good deal of sensitivity. Though Helmholtz's private thoughts remain unknown, ministry officials in Berlin wanted to be sure that Boltzmann was not left adrift. Schulze's communications with Boltzmann's wife made it clear to them that, as Helmholtz's wife had guessed, Boltzmann would not "fit in." If Berlin might be better off without Boltzmann, so too Boltzmann was probably better off in Austria. But before letting Boltzmann go, they made sure he was in a position to continue the same existence in Graz that he had apparently been comfortable in for many years.

Boltzmann himself, however, refused to fall in quietly with this conclusion. Only a week after being told that Berlin was releasing him from any prior obligation, so that he no longer had to make the wrenching choice between Graz and Berlin, Boltzmann was writing yet again to inquire if the decision was really final. He said his request to be let off the Berlin appointment had been made "in the greatest agitation." Since then he had been troubled "night and day . . . by the bitterest regret." He asked for a telegram in return bearing simply the words "too late" or else "still time." Communicating again through Schulze, the Berlin ministry let Boltzmann know that the decision was indeed final and that nothing was to be gained by further inquiry. Finally, at the end of July, Boltzmann formally declared to Vienna his intention to remain in Graz, asking at the same time for permission to take time off until the middle of September.

Still this was not quite the end. Over the summer he gained a little calm, and on returning to his teaching duties found them more onerous than ever. In October he wrote once more to a Berlin official, letting him know, as if to demonstrate his own worth, that the Austrian ministry had given him a pay raise, and asking if there was any chance of regaining the "forfeited" position under the previously negotiated conditions. The minister replied with a good deal of patience, explaining that circumstances had moved on and that, in his opinion, Boltzmann would be best advised to hold on to what he had.

Boltzmann made one last try. In December he sent Helmholtz a new scientific paper, which he admitted was a minor piece, but which, he hoped, would provide some evidence that he had recovered from his "overstrained nerves." He regretted that his weakness caused him to pass up a chance that in normal circumstances he would not have let slip, declared that he was still eager to find a more suitable position, and let Helmholtz know how grateful he would be for any help the older man might be able to provide in that respect. Helmholtz's reply–if indeed there was one–has not survived, as indeed have none of the letters from Helmholtz to Boltzmann. And that, a year after it began, was the final installment of Boltzmann's unhappy tangle with Berlin.

The story that Boltzmann had been offered the Berlin position but hadn't taken it became common currency in scientific circles. The reasons and circumstances remained mysterious, however, and in subsequent years Boltzmann himself did nothing to dispel the mystery. More than 15 years later, writing about a visit to the United States, he offered a slender commentary. An American colleague, he reported, had told him that Berlin's reputation was by then falling and that it would have been a good thing for the university if he, Boltzmann, had taken the position there. Boltzmann's modest rejoinder was that his lecturing would hardly have made a great difference, but he observed that a single individual with the right character and energy can influence an institution's destiny by guiding research and appointing the right kind of people. And then he added a cryptic remark: "Many [people] who were not to be had could have been had, if they had been approached in the right way." Boltzmann was never introspective, never one to ruminate on his own failings or failures. It was his own fear and indecision, as inexplicable to him as it was to anyone else, that had caused him to back away from Berlin, but many years later he had apparently convinced himself it was all Berlin's fault, for not dealing with him as he should have been dealt with.

After the Berlin appointment had finally slipped his uncertain grasp, Boltzmann could no longer be easy in Graz. His assistant von Ettingshausen, who had taken care of all the administrative duties that Boltzmann had no time for, was gone. The Austrian authorities had needed to come up with only a modest increase in salary to obtain his unhappy commitment to stay in Graz. In his aspiration toward the Berlin position, Boltzmann seems to have felt, for the first time, a sense of his own possible inadequacy and the potential for failure. Yet Berlin had, after all, thought him worthy of the position. It became common knowledge that Boltzmann was looking for a new home.

That year brought further unsettling death. In March his oldest son, Ludwig, died at the age of 10 from infection due to appendicitis. Boltzmann was additionally distraught because the seriousness of his son's illness had escaped him; one doctor had misdiagnosed the problem before another had seen, too late, the real cause. The following year his sister, Hedwig, died in her early forties of unrecorded causes. His desire to leave Graz became more urgent still.

Any concern over Boltzmann's erratic behavior–the details of which were in any case not known in full to many people–was exceeded by the desire of other institutions to add so prestigious a physicist to their ranks. In the end, Boltzmann heeded a call from Munich, although not until the university there had assembled what Boltzmann deemed an acceptable remuneration by putting together the salaries of two recently deceased professors.

His actions over the previous two years had caused perturbation and upset in Graz, as well as elsewhere, but the university naturally held a party to mark his departure and celebrate his new appointment. Either from embarrassment over the trouble he had caused, or else in perfect obliviousness, Boltzmann made a speech that began with a startling display of false modesty. "When I learned, a few days ago, that today's ceremony was being planned, it was at first my firm intention to ask you to desist. For how, I asked myself, can one man deserve such an honor? We are all but collaborators in a great enterprise, and everyone who serves his position dutifully deserves equal praise." If the Berlin debacle had cracked his self-confidence a little, the fissures had evidently closed up again.

In Munich, Boltzmann was given his own institute of theoretical physics, independent of experimental physics. In the fall of 1890, 14 years after his arrival in Graz, he was settled in Munich with his wife, surviving son, and two daughters. He was then 46 years old.

CHAPTER 6

The British Engagement

Parsons, Lawyers, and Physicists

BOLTZMANN'S WORK BECAME WELL KNOWN in Great Britain from an early time. In 1876, Henry William Watson, a Cambridge graduate, published a short book called *A Treatise on the Kinetic Theory of Gases*. In his introduction, Watson credited Daniel Bernoulli for conceiving the essential idea almost a century and a half earlier, but he acknowledged Maxwell, Clausius, and Boltzmann for developing the theory in a genuinely scientific way. The purpose of his *Treatise* was to translate and amplify, where necessary, the arguments and theories that Boltzmann had devised, particularly in the famous 1872 paper that presented what became known as the H-theorem.

Watson observed in his introduction that he "felt some difficulty following Dr. Boltzmann's reasoning" in his efforts to deduce the second law of thermodynamics (the rule of increasing entropy) from the kinetic theory. He tried to substitute his own version of some of the arguments but ran into trouble again; he then consulted "his friend Mr S. H. Burbury, formerly fellow of St. John's College, Cambridge" in an attempt to iron out the difficulties. Between them, they came up with a modified version of some of

Boltzmann's arguments, which they thought dealt more successfully with the question of inferring inexorable changes in the properties of a volume of gas from the microscopic motions of constituent atoms.

Although Maxwell died just three years after the appearance of Watson's small book, his influence persisted. From Cambridge came a new generation of younger physicists who were profoundly interested in kinetic and atomic theory and who carried on a lively debate in Britain over the next two or three decades. Watson and Burbury belonged to that now-vanished breed, the gentleman-scholar of science. Neither made their living as professional scientists. Watson entered the church and Burbury spent most of his life as a London lawyer, but both, over the years, continued to make useful incursions into the most abstruse areas of mathematical physics. In 1885, they collaborated on a book, *The Mathematical Theory of Electricity and Magnetism,* and their names showed up regularly in the pages of *Nature,* the weekly scientific journal founded in London in 1869 and still thriving today.

At the end of the introduction to his *Treatise on the Kinetic Theory,* Watson signed himself from Berkswell Rectory in Coventry. Throughout the 19th century, the parsonages of the Church of England provided many a safe refuge for gentleman-scholars of all interests. Charles Darwin himself may well have become a country vicar, except that the opportunity to travel to South America on the HMS *Beagle* came up at an opportune moment.

Centuries earlier Lucretius had taken the idea of atoms as a foundation for atheism, and as modern kinetic theory developed, the hint of atheism, for anyone who wished to think about it, remained strong. It was, after all, simply the random, strictly mechanistic motion and collision of atoms that endowed gases with their tangible properties, and if kinetic theory were taken seriously, it implied that once the atoms had been set into motion, their subsequent behavior should be entirely determined by Newton's laws of mechanics. The French physicist and mathematician Pierre Simon de Laplace made the essential point in 1814: to an intelligence that knew the state of motion of every piece of the world and the physical laws that governed them, "nothing would be uncertain, and the future, like the past, would be before its eyes." The philosopher Roger Boscovich (a Catholic priest) had, however, made a similar observation some 50 years earlier. "If the law of forces were known, and the position, velocity, and direction of all the points at any given instant, it would be possible . . . to foresee all the necessary subsequent motions and states, and to predict all the phenomena that necessarily followed from them."

Such considerations set up a potential battleground between physics and religion. If mechanics and the blind, random motion of atoms were all, where was the hand of God? This does not seem to have deterred men such as Watson from writing their sermons for Sunday and pondering kinetic theory in their spare hours. English country parsons have never been renowned for the depths of their theological inquiries, and Maxwell's school at Cambridge was in any case profoundly pragmatic. He and his students wanted to understand how kinetic theory worked and what it implied. Philosophical debates, whether over the existence of God or the existence of atoms, were not their style. Maxwell himself was religious in a stern Scottish way (his occasional letters to his wife, full of solemn exhortations to ponder the drearier parts of the New Testament, contrast oddly with his lively and sometimes acidulous letters to Tait and others), but he kept science and religion separate. Mechanics and kinetic theory still left room for a creator–a divine agency who would set all the atoms of the universe in motion at some primordial moment–and that lacuna in physical theory seems to have provided enough room for Maxwell and many others to reconcile God and reason.

There was, in any case, a purely scientific question that cropped up in considering the connection between kinetic theory and determinism. Although Boltzmann's famous theorem said, apparently, that the mathematical quantity H, derived solely from atomic motions, must always decrease, Maxwell, Thomson, and Loschmidt had all realized that this could not be strictly true. For every set of atomic motions that would cause H to decrease, there must be another, the time-reversal of the first, that would cause H to increase. Since H was entropy with a minus sign in front of it, such changes would correspond to entropy decreases, contrary to the second law. This seeming contradiction led to a simmering puzzlement for many years in England, but Maxwell's early death left the debate leaderless for some time.

A new round of attacks on the problem began in the 1890s. Edward P. Culverwell, a Dublin physicist, published a short comment in the *Philosophical Magazine* of July 1890 to the effect that it seemed impossible to prove by purely mechanical means that a collection of atoms or molecules must necessarily evolve toward thermal equilibrium. After noting that Boltzmann had claimed that kinetic theory could predict just that, he stated yet again the reversibility objection: "for every configuration [of atomic motions] which tends to an equal distribution of energy, there is another which tends to an unequal distribution." Culverwell concluded that kinetic theory alone could never succeed in explaining the laws of thermodynamics, and he proposed instead that the molecules of a gas must move about in some sort of "ether"–a hypothetical and more or less intangible fluid, reminiscent of the "caloric" of old, with which molecules could exchange energy back and forth. Even then, he went on to say, the problem of explaining thermodynamics had not been resolved, only shunted aside into the unknown properties of the mysterious ether.

The burgeoning debate in Britain and Ireland came to Boltzmann's attention. In 1892, he traveled to Dublin for the 300th anniversary of Trinity College and met his critic Culverwell. Two years later he was in Oxford, to receive an honorary degree and also to attend the annual meeting of the British Association for the Advancement of Science, an organization encompassing scientists and philosophers of all interests. The association had convened an effort some years earlier to try to resolve the mysteries of kinetic theory and thermodynamics, prompted in large part by Culverwell's remarks of 1890 and a longer analysis from Burbury, published in the *Philosophical Magazine* later that same year. Doubt hung over not only the validity of Boltzmann's theories but also the truth of kinetic theory itself. Did a gas consist of atoms or molecules alone, or was an ether necessary too?

The 1894 British Association meeting in Oxford was a momentous occasion for Boltzmann. He met with something he had never really encountered in the German world of physics: friendly criticism. For the most part, the British physicists were avowed atomists, or at least interested in atomic theory in a neutral and pragmatic way. They could see difficulties, and they continued to puzzle over Boltzmann's arguments, but they wanted to understand the theory better, not use the difficulties as a reason to dismiss the whole idea. "The part which Prof. Boltzmann took in these discussions will long be remembered," one of the attending English physicists recalled years later. He was flattered and amused at receiving his honorary degree, which made him a Doctor of Laws; "It were better they made me Doctor of Science," he said, but his English colleagues assured him the title was appropriate, since he was, after all, in Oxford in his capacity as an expert on the laws of thermodynamics.

By the time of his visit to Oxford, Boltzmann had been in Munich for over three years, and he appeared to be thriving. He confined his teaching mainly to a few high-level graduate seminars, which were typically attended by only half a dozen students or so. In these circumstances, his best qualities stood out. He was conscientious and thoughtful, determined to impart understanding. A Japanese student who visited Europe in the early 1890s reported that Boltzmann was "an odd little fellow" but that in his lectures he "is gentle and honest, and has a personality to be loved by everybody, rather in contrast to his features." Boltzmann was by this time increasingly rotund, and his bulky frame, thick eyeglasses, heavy beard, wild hair, and incongruously high-pitched voice combined to make him the very picture of the unworldly scientist.

Boltzmann's life in Munich had many gratifications. Although the family moved a number of times, apparently at Henriette's urging, they generally lived near the center of the city and were able to frequent the theater and the opera. Boltzmann could get his fill of Wagner, and they attended the first performance of an opera by Henriette's childhood friend Wilhelm Kienzl. In 1891, his third daughter, Elsa, was born. His teaching and administrative duties were never onerous, and he was often able to shed the more tiresome responsibilities. Such were the benefits of being a sought-after academic commodity. And as he had not in Graz, he found an agreeable collection of physicists and mathematicians in Munich who got together once a week as the "Hofbräuhausgesellschaft"–the brewhouse bunch, one might say. More than a century later, the Hofbräuhaus remains a prosperous Munich drinking establishment, catering mainly to tourists but perhaps still allowing room for occasional scholarly debate.

If Boltzmann personally was thriving, kinetic theory in Germany and Austria was not doing so well. A new voice had gained prominence among the critics. After the debacle over Berlin, the position that Boltzmann had let slip was finally awarded to a young theoretical physicist named Max Planck, who was then just 32 years old and who was to remain in Berlin until his death in 1947. Planck was born in 1858 in Kiel, on the North Sea, but his early schooling came in Munich, where his father was appointed a law professor at the university. Planck attributed his fascination with science to a moment of epiphany. When an enthusiastic schoolteacher introduced the bright young student to some elementary ideas in physics, Planck later recalled, "My mind absorbed avidly, like a revelation, the first law I knew to possess absolute, universal validity–the principle of the conservation of energy." It was the precision and the inviolability of laws such as these that impressed Planck. He saw how "pure reasoning can enable man to gain an insight into the mechanisms of [the world]." Some years later he experienced again the same sense of revelation when he discovered Clausius's precise formulation of the two laws of thermodynamics. The quest for absolute laws and unshakable truth became a mainstay of Planck's thinking about physics.

He studied for a year in Berlin, under Helmholtz and Kirchhoff, but the teaching of these great men, he recalled, "netted me no perceptible gain." Helmholtz came to his lectures ill-prepared, and confused his listeners. Kirchhoff prepared fastidiously, and bored them.

Returning to Munich, Planck completed a Ph.D. thesis on the laws of thermodynamics, and he continued for the next few years in a substantial effort to refine and extend Clausius's work so as to express thermodynamics laws as precisely and comprehensively as possible. Much of this work was, according to Planck, "widely ignored" at the time. Quite as much as Boltzmann, Planck had a capacity to see himself as an unappreciated laborer in the mines of science, a character trait that may have amplified the enmity that developed between them.

Even so, Planck proceeded nicely up the academic ladder, landing the Berlin position as a young man. Now working alongside Helmholtz rather than attending his lectures, Planck came to see the older man as "the very incarnation of the dignity and probity of science."

Throughout the first half of his scientific life, Planck held fast to the notion that scientific laws must embody some element of absolute and inviolable truth. And since he was enamored above all of the laws of thermodynamics, he was bound to dislike Boltzmann's perspective. He saw that kinetic theory could not produce certainties in the thermodynamic laws, only probabilities and tendencies, and he took this as a fundamental failing. In a paper published in 1882, he wrote that "the second law of the mechanical theory of heat is incompatible with the assumption of finite atoms. . . . a variety of present signs seems to me to indicate that atomic theory, despite its great successes, will ultimately have to be abandoned."

A decade later, his opinion had not changed. Speaking in 1891 to a meeting of the German Society of Natural Scientists in the city of Halle, Planck said of kinetic theory that "the remarkable physical insight and mathematical skill exhibited in conquering these problems is inadequately rewarded by the fruitfulness of the results gained." Since the physical insight and mathematical skill Planck mentioned were largely those of Boltzmann, this was a sharp dig that all his talents and effort were going into a project that, Planck believed, could not possibly succeed. If kinetic theory and atomism predicted that the laws of thermodynamics were mere approximations, then it was so much the worse for atomism.

A few years later, Planck served as the editor of a posthumous collection of Kirchhoff's work and used his introduction to, as Boltzmann saw it anyway, once again cast aspersions on the value of kinetic theory. For some years, Planck continued to maintain the attitude that physical reasoning based on probability was a stopgap measure at best. It wasn't that Planck found the idea of atoms philosophically objectionable in any fundamental way, simply that as a hypothesis it led to predictions he thought unsound, inconsistent, and undesirable.

And he had a point. Boltzmann's response to the reversibility objection, when Loschmidt had first raised it in 1876, was that admittedly atomic motions might in some circumstances conspire to decrease entropy rather than increase it. But on the whole, systems would evolve from less probable to more probable states, and all would be well. Planck's objection to this was twofold. First, saying that systems would evolve from lower to higher probability was, after all, only a matter of probability; it didn't rule out the possibility that systems could occasionally go the other way, and that was a point Boltzmann seemed slow to grasp. Second, was it not, in any case, begging the question to assert that systems must begin in states of low probability? It was all very well to analyze possible states and assign them probabilities, but, as Planck forcefully pointed out, the evolution of any one particular state to another "is determined not by probability but by mechanics." Given an initial state, in other words, its next state was in principle fully and absolutely determined, if only all the atomic motions were exactly known. It was in the jump from specific states of actual gases to probabilities of states of hypothetical gases that Boltzmann, in Planck's view, was tripping up.

In the early 1890s, in Germany, kinetic theory received little attention, and the tenor of Planck's comments typified what little there was: acute, but dismissive. In England Boltzmann found a different attitude. The criticism was equally acute, if not more so, but the apparent flaws of kinetic theory were taken as an invitation to understand the subject better, not to throw it overboard. Boltzmann (although he was admittedly working in a foreign language) had a little difficulty adapting his customary rhetoric to these more benign circumstances.

Following the 1894 British Association meeting, a flurry of letters and notes appeared in *Nature*. Culverwell, the Irish physicist, opened the correspondence with a plaintive lament. In a bare half-column note published in October, he stated yet again the reversibility objection and asked, "Will some one say exactly what the H-theorem proves?" Burbury, the following month, rose to the challenge. After briefly correcting a technical argument of Culverwell's, he makes clear that in his view the H-theorem concerns probabilities: "there is a general tendency for H to diminish, although it may conceivably increase in particular cases." To which he adds the not altogether relevant observation that "just as in matters political, change for the better is possible, but the tendency is for all change to be from bad to worse."

Boltzmann's reply to his English critics and commentators came in February 1895, in a three-page communication in *Nature*. In their short notes, Culverwell and Burbury had straightforwardly stated and discussed some technical matters in Boltzmann's work. By way of reply, Boltzmann offered a dense disquisition which started off with a lofty declaration of purpose. "I propose to answer two questions," he intoned. "(1) Is the Theory of Gases a true physical theory as valuable as any other theory? (2) What can we demand from any physical theory?" There followed several paragraphs of musings on the nature of physics as opposed to metaphysics, and on the extent to which the success of theoretical explanations allows one to decide the truth of the assumptions on which the theory is based. Boltzmann's manner is strikingly different from the tone of the direct and essentially simple questions that had been put to him. His opening paragraphs are a swirl of high-minded philosophical generalities and somewhat arch statements about the nature and purpose of scientific theory. Such matters were not, to put it mildly, of any great concern to Culverwell, Burbury, and the rest, who merely wanted to know how kinetic theory worked, not how its intellectual status stacked up in some sort of philosophical contest between other theories and forms of knowledge generally. His remarks do at least give the English reader some idea of why so many German scientists (and Maxwell) professed to find Boltzmann hard to follow.

He does, nevertheless, get to the point. The kinetic theory, he eventually says, "agrees in so many respects with the facts, that we can hardly doubt that in gases certain entities, the number and size of which can roughly be determined, fly about pell-mell." In short, Boltzmann thinks atoms and molecules exist, and with that cleared up, he addresses what Culverwell and Burbury have to say.

Culverwell's objections, Boltzmann says, "bear the closest objection to what I pointed out" in his 1877 response to Loschmidt's reversibility question. "There I pointed out that my Minimum Theorem [the H-theorem], as well as the so-called second law of thermodynamics, are only theorems of probability." Culverwell is merely observing, as Loschmidt did, that some atomic motions must cause H to increase. Boltzmann's response is that the mere possibility of such motions does not mean they have any large overall importance, as long as they are unlikely compared to motions that make H decrease, as the theorem requires. He then elaborates on this question of probability in order to answer Burbury's more sophisticated criticisms.

The upshot of all this is that although H may "increase or decrease, the probability that it decreases is always greater." And this, according to Boltzmann in 1895, is what he said in 1872, said again in 1877, and indeed has been saying all along.

Here Boltzmann is shading the truth, or at least proffering a somewhat partial account of his own previous scientific statements. In the 1872 paper proposing the H-theorem, it seems quite clear that Boltzmann thought he had established an absolute rule: H must go strictly in one direction. In light of the reversibility objections, he changed his tune a little and began to say he had proved only that H would almost always decrease, with high enough probability that the opposite cases could be neglected. It may be that in Boltzmann's mind, once these objections came along, he revised his own estimation of what he had done and decided that was indeed what he had meant in the first place. But to his critics this was an implicit acknowledgment that the kinetic theory was something of a ruse, if its supposed predictions slipped a little from one paper to the next.

Even in later years this contradiction in Boltzmann's published works was not fully resolved. In his monumental two-volume monograph *Lectures on Gas Theory,* first published in 1896 and 1898, both interpretations of the H-theorem can be found. Sometimes it says that H strictly decreases, but elsewhere Boltzmann acknowledges that movements in the opposite direction are possible, if not very likely. In later editions of *Gas Theory* the more contentious statements about the H-theorem were simply dropped, without explanation.

Despite Boltzmann's protestation, Culverwell's original request–"Will some one say exactly what the H-theorem proves?"–still received no wholly unambiguous answer. By this time, however, it was clear that Boltzmann was thoroughly convinced of the probabilistic nature of the second law, and was furthermore equipped with some quantitative reasoning that explained why departures from the law of increasing entropy were bound to be negligible in all realistic circumstances. That, it seemed, was the best that could be hoped for. Boltzmann's reply, even to his collegial critics in England and Ireland, betrayed a certain impatience. "Hence," he wrote in *Nature,* "Mr. Burbury is wrong, if he concedes that H increases in as many cases as it decreases, and Mr. Culverwell is also wrong, if he says that all that any proof can show is that [H will decrease in an average sense]."

THIS LONGISH communication appeared in *Nature* in February 1895, and at its conclusion Boltzmann signed himself from the

Imperial University of Vienna. He was no longer in Munich. Though he appeared to be happy there, he was struck by homesickness from time to time. In October 1892, he wrote after a gap of many years to his old friend Josef Loschmidt, mainly to see if Loschmidt had a copy of a book he had written that was now out of print. To introduce himself after such a long intermission he began by saying, "our lives have traveled so far apart that I must first give you the news that I am still living, though indeed no better here than in dear old Austria."

Loschmidt wrote back to say that his health was no longer strong, and Boltzmann, not one to be outdone in complaints over poor health, replied to commiserate and to tell Loschmidt of his own problems in that regard. "I have often noticed," he told Loschmidt, "that a delicate constitution is hardier and more durable than a robust one." And he added a melancholy comment he recalled from his friend Schulze, who had acted as the go-between in the Berlin affair: "Once when I was speaking warmly to him of my plans and dreams for the future, he looked at me sharply and replied: don't hope for too much, believe you me, it always gets worse. I feel that way more and more." What prompted this gloomy observation is not clear, unless it was the realization that both he and Loschmidt were getting old and in unreliable health.

In the meantime, the loss of so prestigious a scientist–his status further confirmed by the honorary degree from Oxford–was a continuing irritation to the Austrian court. Officials in Vienna were constantly alert for any opportunity to bring the hero home. The first chance came in January 1893, when Josef Stefan, Boltzmann's mentor and the longtime director of the Institute of Physics in Vienna, died at the age of 57. Boltzmann was an obvious choice for the open position, but not the only one. A faction within the university preferred Ernst Mach, still in Prague, whose reputation was growing. He was a good teacher, and over the previous decade his large books on the history, meaning, and philosophical development of science had begun to attract a following. Mach was actively interested not only in physics but in psychology and chemistry, and he had the kind of intellect that yearns to construct all-encompassing systems, tying all of science into one coherent structure. Boltzmann might be the more profound scientist, but his greatest achievements concerned essentially one problem–and one that, moreover, was now largely underestimated or even discounted in Germany and Austria. Among the general faculty of the University of Vienna were many professors who saw Mach as a greater asset than Boltzmann, and with some reason.

Boltzmann, approached informally, made it plain he would be interested to hear an offer from Vienna. He praised Stefan's memory and the institution he had created, and declared it would be "his ideal" to teach at such a place. The recipient of this letter came away with the clear and unmistakable impression that Boltzmann was eager to return home, to his own country and to the city of his birth. An offer was duly made, despite some protestation from the Mach camp. Boltzmann turned it down. Even after the Berlin debacle, Boltzmann was apparently ready to play one university against the other. The authorities in Munich pushed an older physicist into retirement, hired a new young physicist to take care of the elementary lecturing in theoretical physics (which Boltzmann found increasingly irksome), and gave Boltzmann himself a considerable salary increase on the understanding, so the Bavarian authorities believed, that he would stick around for a while.

Back in Vienna, Boltzmann's rebuff caused consternation. In the end, they reorganized the physics faculty, shuffled around some of the existing professors, and after a few months began looking to hire a professor of mechanics. The names of Mach and one or two others came up.

But in June 1893, informal word came again from Boltzmann to the effect that although he had taken the Munich offer, he personally felt he was under no obligation to stay there beyond the end of the 1893–94 academic year. If he were to leave Munich then, "no one could say a word against me." He might still be interested in Vienna, he intimated.

In Vienna, an internal ministry document sought to reassure doubters that Boltzmann really did want to return to his native land. His departure from Graz to Munich was put down to mental disarrangement over the loss of his son; it was even hinted that a change of scene had been strongly recommended by a doctor. With these sorts of assurances overcoming any doubts that might have been prompted by Boltzmann's behavior just four months earlier, the ministry sent him a generous offer at the end of 1893. Boltzmann's reply was to ask for more money, both for salary and for his researches. Also, at his wife's urging, he insisted on an appropriate pension arrangement, since in Munich he had none.

The Austrian ministry evidently decided that it was prepared to pay any price to bring Boltzmann home. They gave him what he wanted, and he agreed to take up his new position in Vienna beginning in the fall of 1894. He had now only to escape from his obligations at Munich, which he did by declaring that the second offer from Vienna, the one made in June, was an entirely different matter from the January offer, since in the meantime one of the older Viennese physicists had died, opening up a quite different vacancy. It was true that one Gottlieb Adler, a physicist, had died, but the interpretation of this event was Boltzmann's. He did, however, keep to his word, and he returned to Vienna in September 1894 as professor of theoretical physics. At the age of 50, he had attained the most senior position for a physicist in the city of his birth, becoming director of the fondly remembered institute where, some three decades earlier, his youthful brilliance had first shone and his life as a scientist had begun with unclouded promise.

CHAPTER 7

"It's Easy to Mistake a Great Stupidity for a Great Discovery"

Philosophy Seduces Physics

BOLTZMANN WAS INVIGORATED BY his lively engagement with the English physicists in Oxford over the meaning of kinetic theory. In September 1895, the German Society of Natural Scientists was scheduled to meet in the Baltic port city of Lübeck, and in June of that year Boltzmann wrote to his friend the chemist Wilhelm Ostwald that he "would like to bring about, if possible, a debate a la british association [sic], mainly for my own education. It is essential that the chief leaders of opinion attend. I hardly need tell you how important to me in particular your own attendance would be."

Boltzmann had met Ostwald in 1887 when the younger man, then 34 years old, had come to Graz for a few months to study. Ostwald quickly made a name for himself as one of the greatest of German chemists, a prominent leader and organizer, as well as an original scientist. He largely founded the discipline that is now called physical chemistry, in essence a marriage of chemistry and thermodynamics. The aim of physical chemistry is to understand the exchanges of energy involved in chemical reactions, the influence of temperature and other external conditions on reaction rates, and in general to elucidate the dependence of chemical changes on physical circumstances.

Boltzmann had been impressed by Ostwald, who had in turn enjoyed the friendly reception he found in Graz. At the time of his brief sojourn in Graz, Ostwald was in the process of moving from the University of Riga (now the capital of Latvia), which was his birthplace, to the old and important University of Leipzig, 100 miles southwest of Berlin in the German state of Saxony. There he stayed for almost two decades, building up an enormously influential department. He was the founder in 1889 of the *Zeitschrift für Physikalische Chemie,* the first scientific journal for his discipline.

Despite his early encounter with Boltzmann and kinetic theory, Ostwald fell under the influence of Mach and his antipathy toward theorizing. It was possible at that time to be a chemist without believing that atoms and molecules were real objects. Rather, they could be viewed as convenient but abstract accounting devices, notional divisions of matter, whose essential purpose was to allow a chemist to keep track of the bookkeeping in chemical reactions. Hydrogen and oxygen were well known to combine in two-to-one proportions to form water, but that did not necessarily mean that two bona fide bits of hydrogen linked up to a single bit of oxygen. In chemistry, even toward the end of the 19th century, belief in the reality of atoms was optional. Many chemists found it not even an interesting question, too abstract for their tastes.

Ostwald, however, had a strong philosophical bent, and this kind of agnosticism proved unsatisfactory. His knowledge of physics, moreover, made him hanker after a version of chemistry based on some fundamental principles. Atoms wouldn't do, because they constituted a form of abstract theorizing–metaphysics, in Mach's dismissive vocabulary. Instead, Ostwald became enamored of an idea called energetics or energeticism, which was built on the notion that energy, since it is an observable and tangible thing, ought to constitute the elemental stuff of scientific explanation. In this view, heat was undoubtedly a form of energy, but as to the nature of heat, no more could or should be said. The law of energy conservation was the prime rule, and other laws, including other laws of thermodynamics and even Newtonian mechanics, ought to flow from that essential principle.

To some energeticists, atoms were a needless but not necessarily sinful hypothesis. But in the quest for philosophical purity, the leaders of this new movement developed a positive antipathy toward atomism. In 1887, the chemist Georg Helm published a book called *The Theory of Energy,* which argued for the recognition of energy as the fundamental stuff of the physical world. Helm's book had little immediate impact, but its message was seized on by Ostwald: this was the foundation he needed for his interpretation of physics and chemistry. During the early 1890s, Ostwald energetically took up the cause of energeticism, hoping that somehow all the known laws of physics could be shown to derive from rules governing the transformation of energy. There would then be no talk of atoms. In 1892, Ostwald visited Boltzmann briefly in Munich, and Boltzmann wrote shortly afterward asking that if he and Helm were ever to get around to formulating their ideas in a more systematic way, they should please keep him informed of progress.

A little later Ostwald indeed sent a manuscript, which he described as a sketch of the foundations of energeticism. He displayed a certain awkwardness and trepidation, admitting himself to be a "bungler" in mathematics, asking for Boltzmann's critical analysis but also asking for any comments to be kept between them. Later, thanking Boltzmann for his "friendly opinion" of the manuscript, he added, "you can hardly know how valuable this is to me. In such matters it's easy to mistake a great stupidity for a great discovery."

Boltzmann's replies also show a delicacy of phrasing unusual for him. He emphasizes his admiration for the effort Ostwald is making, and cautions him against taking his criticisms too badly. He warns against excessive rigidity of thinking: "against the dogma that nature can be explained only mechanically (through the motion of atoms) I would not like to set the opposite view, that it cannot be explained in that way." Ostwald admired and liked Boltzmann, but was also fearful of submitting his ideas to what he acknowledged were the other man's great powers of physical and mathematical analysis. Boltzmann in turn liked and admired Ostwald, but at heart thought his crusade for energeticism philosophically dubious and scientifically just plain wrong.

Against this background, the Lübeck debate unfolded. Ostwald and Helm took the field for energeticism, and Boltzmann, seconded by a young mathematician named Felix Klein, argued the case for atoms. The event was, despite Boltzmann's original intent, nothing like the debate at Oxford. There, interested physicists had gathered to discuss kinetic theory and hammer out, if they could, the meaning of the H-theorem, the nature of reversibility, the probabilistic nature of the laws of thermodynamics, and so on. It was a scientific debate devoted to the elucidation of subtle issues implicit in a profound but still unfinished theory.

At Lübeck, on the other hand, Boltzmann and Klein had to defend the very essence of kinetic theory against opponents who simply did not believe in the existence of atoms, who saw Boltzmann's work as elaborate mathematical speculation founded on pure assumption, and who would not even allow Boltzmann the privilege of thinking that his theories constituted a respectable kind of scientific investigation. Ostwald and Helm were there not to debate the merits of kinetic theory but to deny them altogether.

Ostwald, moreover, was something of an orator, where Boltzmann was generally not; he could be a fierce and impressive speaker, but he was not always articulate. Ostwald was fluent, and adept at putting his ideas into easily assimilable form. Over the course of his life he wrote many books, both technical and nontechnical, concerning general issues of philosophy and intellectual development as well as science itself, and he capped it off with a three-volume autobiography.

The debate took place on September 16, 1895, in a large hall filled with hundreds of eager listeners, and it occupied most of the day. Helm and then Ostwald sketched their view of energetics, arguing that although it was still far from complete, it held the promise of all-encompassing explanatory power based on elementary principles. They were, they claimed, offering a program of research, not a finished body of knowledge. Both chemists, they tripped up when they tried to explain how the well-known laws of mechanics and thermodynamics were supposed to follow from the conservation of energy alone.

Boltzmann began his reply amicably enough, making some general remarks about the need to explore a range of scientific hypotheses in order to move science forward. He professed himself eager to avoid hostility: "I hope I can count among my closest friends the scholars whose names I will mention later, and I reckon countless of their achievements among the most outstanding scientific works; I set myself only against their specific publications on energeticism. This observation will suffice to prevent the subsequent attacks directed at any of their conclusions or mathematical formulas from being imbued with the slightest personal character."

Having made this lofty declaration of neutrality, Boltzmann then tore deliberately into the theoretical pretensions of his opponents. What followed was a lengthy and fairly technical analysis, but the gist of it was simple. He explained what any physicist knew to be true. Newtonian mechanics was based on more than just energy conservation. The second law of thermodynamics was distinct from the first and not, as some physicists had originally thought, derivable from it. These were not questions of philosophical preference but matters of physical reasoning and mathematical proof. Energeticism simply could not achieve the things it set out to achieve. He may have included no personal remarks, but Boltzmann made no attempt to sugarcoat his lack of respect for his opponents' scientific aims. He explained in unrelenting detail all the reasons why he thought the cherished hopes of the energeticists were utterly at odds with physics as it was then understood.

Helm, writing to his wife the next day, said that "things went hard." Boltzmann and Klein "touched on things which during my preparation and in my correspondence I had not anticipated, and which seemed inappropriate, and it was hard for me to put in even a few words of clarification." Ostwald, in his autobiography, recalled that he felt surrounded by "closed antagonism," and that the Lübeck debate was "the first time I personally found myself confronted by such a unanimous band of downright adversaries." Svante Arrhenius, the Swedish chemist who had come under Boltzmann's influence at Graz, also attended the debate and later wrote that "the energeticists were thoroughly defeated at every point, above all by Boltzmann, who brilliantly expounded the elements of kinetic theory. . . . Ostwald was quite exhausted when the discussion ended, and Helm spoke of having been lured into an ambush."

Many years later, at a meeting held in wartime Vienna to mark the 1944 centenary of Boltzmann's birth, another physicist recollected the contest. Arnold Sommerfeld, who succeeded Boltzmann in Munich and became, in the early part of the 20th century, a champion of the new quantum theory, had attended the Lübeck meeting as a young man. He compared the debate between the dogged Boltzmann and the nimble Ostwald to "the fight of the bull with the lithe swordsman. But this time, in spite of all his swordsmanship, the toreador was defeated by the bull. The arguments of Boltzmann broke through. At the time, we mathematicians all stood on Boltzmann's side."

By all accounts, in other words, even according to Ostwald and Helm, the much ballyhooed debate in Lübeck was a rout, with Boltzmann on the winning side. The friendship and correspondence between Boltzmann and Ostwald cooled for some years.

Still, though Boltzmann must presumably have felt some sense of victory, he had not succeeded in convincing Helm or Ostwald of the error of their ways. It was a perpetual annoyance to him in his university lecturing if, despite his best efforts, he could not get all the students in the room to grasp what he was saying. Likewise, the fact that he could not persuade his opponents in Lübeck to alter their thinking would have struck Boltzmann as a failure on his part. Boltzmann did not find, in the Lübeck debate, anything like the lively exchange of views he had encountered the previous year in Oxford. He and his opponents had gone at each other stubbornly, stating and restating their opinions, but in the end everyone was left standing and none of the principals had switched sides. If his aim had been not just to explain but to persuade and convert, Boltzmann had failed.

Back in Vienna, moreover, he now found himself face to face with the very symbol of continuing opposition to atomism. Ernst Mach had finally escaped from Prague, and was now a professor in Vienna–and a professor, no less, of philosophy.

The two men had been in Vienna at the same time once before. When Boltzmann was there as an undergraduate, Mach had already graduated and was offering a variety of lecture courses in general physics–none of which Boltzmann attended. When Stefan had taken over the Institute of Physics, Mach had gone off to Graz; Boltzmann went there a couple of years later, but by then Mach had left "cheerful, friendly Graz for beautiful, gloomy Prague."

Frequently riven by nationalistic conflicts between its German and Czech inhabitants, Prague was not always a pleasant home for Mach. His family was in fact of Czech origin, and Mach, to the inhabitants of Prague and to the students and officials of the university there, was a recognizably Czech name. He spoke only German, however. The mere contrast between his name and his language was enough, in Prague, to raise suspicions.

Over the years, Mach made a number of attempts to return from Prague closer to the center of Austria-Hungary. On a few occasions he found himself in competition with Boltzmann, and coming up short. Mach had some desire for the position that Boltzmann took in Graz after Toepler's departure, but his cause had too little support. In the middle 1880s, the university in Munich was looking for an experimental physicist, and Mach's name came up, but at that time he was unable to obtain a release from Prague. A few years after that, it was Boltzmann who landed up in Munich. After Boltzmann's first refusal of the position made vacant in Vienna by Stefan's death in 1893, Mach's name was on the list again, but instead of going to him, the Viennese authorities waited a few months, by which time Boltzmann's renewed availability became known. Boltzmann returned to Vienna in 1894, while Mach, after 27 years, remained in Prague.

Like Boltzmann in Graz, Mach became university rector for a time in Prague and similarly had to deal with fractious students. Prague was a divided city, with German and Czech factions. As elsewhere in the Habsburg Empire, Slavic populations were beginning to coalesce into factions that opposed the rule of Vienna. Count Edward von Taaffe, an Austrian of Irish ancestry, became chief minister to Emperor Franz-Josef in 1879, the same year that Mach became rector in Prague. It was von Taaffe, in one account, who popularized the notion of *fortwürsteln,* muddling through. He was pragmatic and not at all ideological, and he tried to deal with simmering and perpetual nationalistic disputes by creating a system in which all parties felt they had some power and participation. This was a generally well-meant philosophy, but it turned inevitably into a succession of stopgap changes that left no one happy for very long.

In Prague, as a result of von Taaffe's initiatives, the university split into two supposedly equal parts, one Czech and the other German. Neither side wanted to abandon its claims to the ancient buildings of Charles-Ferdinand University, so instead the old institution was divided up, leaving the Czech and German schools sharing the same buildings as reluctant and untrusting neighbors.

Mach, as rector, tried to deal evenhandedly with the factions and was for that reason mistrusted equally by both sides. At a meeting of the German students' association in 1880, he spoke for tolerance and moderation, but those words were forgotten in comparison to strident noises from other speakers, and when riots between the student groups developed over the next few days, Mach was lumped by the Czech press into the pro-German movement. A few years later Mach was rector for a second term, and agitation between the Czech and German halves broke out once again. This time Mach more actively opposed Czech-inspired attempts to reorganize parts of the university and was branded a pro-German. On the other hand, as overt anti-Semitism began to infect the German faction, Mach distinguished himself by defending the Jews and opposing August Rohling, who had been named professor of Hebrew Antiquities and used his position to broadcast ancient blood libels against the Jews. Mach, meanwhile, was attacked with allegations of atheism, which were more or less true, except that Mach remained formally a Catholic and was not one to advertise his irreligious views. In 1884, beset by endless squabbles over university administration, ethnic factionalism, and religion, Mach threw in the towel and resigned his rectorship, hoping to keep his head down and return to his scientific and philosophical writings. As the 1890s approached, the number of German students in physics fell and financial support for his researches waned. Irked by skirmishing between his old, alcoholic technical assistant and a new, incompetent one, Mach put his oldest son, Ludwig, in charge of his lab, which only added to the resentments.

Nationalistic discontent was rising too in the heart of Austria, but still, Vienna was calmer than the fringes of the empire. And the University of Vienna was still the apex of the Austrian intellectual world. Mach had lost out to Boltzmann once again after Stefan's death, but his yearning to be in Vienna was powerful. On top of all his other troubles, Mach's younger son, Heinrich, had killed himself, at the age of 20, a few days after receiving his doctorate in chemistry from the University of Göttingen in Germany. The reasons for his suicide were obscure. Just a couple of weeks later, Mach was in Vienna giving a philosophical lecture on the meaninglessness of cause and effect. His argument, that while one may often observe a certain phenomenon happening after some other, one should resist the temptation to infer any causative link between them, impressed the philosophical faculty and others at the university. A movement began to draw Mach to Vienna as a professor of philosophy rather than physics. In May 1895, he was formally appointed to the chair of the History and Philosophy of the Inductive Sciences. Before leaving Prague he wrote to Boltzmann, saying that he looked forward to a collegial relationship despite their obvious differences of opinion. Boltzmann wrote back in a similar vein, expressing the hope that he would be able to learn from his new faculty colleague.

Mach's philosophy, no matter how elaborate it became, always rested on a simple principle: science should be based on observable facts, not hypotheses or theories. This may seem an innocent enough statement, and may even seem self-evident, as if science were by definition a matter of dealing with facts about the world, but it soon runs into trouble. The exertion of pressure by gas on the walls of a container is, by Mach's standards, an acceptable fact, since it is something that all can agree on. To explain that pressure in terms of underlying atomic motions is an unacceptable hypothesis, since it relies on entities–atoms–that cannot be seen.

Similarly, in Mach's view, heat is a primary phenomenon, a fact, since anyone can feel the difference between a hot dinner plate and a cold doorknob on an icy day. Portraying heat as atomic motion is unacceptable.

Even the most ardent atomists had to accept that the objects of their affections and desires could not be directly observed. The philosophical question, at its simplest, therefore boiled down to a debate over the extent to which it was permissible in science to make hypotheses in order to achieve a fuller or more unified understanding. The atomists claimed that they could explain both heat and pressure in similar terms–atomic motions–so that the hypothesis of atoms led to a deeper understanding.

In Mach's estimation, this supposed increase in understanding was an illusion, purchased at the expense of dreaming into existence particles whose reality could not independently be judged. This argument, conceived by Mach at a time when theoretical physics was young, reverberates still today. During the 20th century, physicists have predicted the existence of many a subatomic particle that was only later proved experimentally to exist, and now some physicists argue for the existence of superstrings and other curious entities that will never be seen directly. It remains, even now, a profound question whether the cost of proposing such very hypothetical objects as superstrings is sufficiently compensated by the benefit in understanding that the hypothesis brings. Mach's critical attitude retains merit.

In his own day, however, Mach sought to deal with his concerns by stifling theory altogether. He insisted that the atomic hypothesis was beyond the bounds of true science, and that the physicist should instead deal with temperature and pressure as fundamental entities in themselves. A scientific understanding of the behavior of gases then amounted to describing empirical relationships between temperature and pressure. Mach's stringent philosophy makes science, in essence, a matter of depiction rather than anything that Boltzmann or his sympathizers would call understanding. As he pursued in his single-minded fashion the consequences of what he took to be an unarguable assumption about the nature of science, Mach came to disparage theories altogether as mere props to understanding. If the kinetic theory of gases was of some assistance in puzzling out new aspects of the measured behavior of gases, then it was not altogether useless or inadmissible, but the point was to eventually disencumber oneself of the theory once new relationships among measured properties were found and established.

Mach began to think of himself as an "anti-philosopher." Any philosophy, any systematization of knowledge, was usually founded on some assumption, whereas Mach, so he claimed, began with no assumptions at all. Stick to verifiable facts, he declared, make the goal of science the finding of mathematical relationships between those facts, and what would result would be an utterly reliable account of the workings of the world. Theories had meaning only insofar as they were useful and practical; the intellectual content of theory had no meaning at all. Or rather, any meaning it had was put there by theorists, not derived from the world itself. Atoms were a fiction, Mach maintained, possibly a useful device but nothing more.

But Mach's own philosophy, whatever he may have thought, had assumptions of its own–principally that facts were unarguable, and that everyone could agree on what the facts were. The stringent belief that one must take every observation at face value is the central unexplained element of Mach's view of the world, and he seems to have held it since childhood. As a young boy, he later recollected, he found great difficulty in understanding why a long table looks wider at the near end than the far end, and he seems in a sense never to have got over this difficulty. He objected to the way artists used tricks of perspective to portray three-dimensional objects on a two-dimensional canvas, as if they were somehow fooling the viewer into accepting a distorted view of reality as the real thing.

His later philosophy amounted to an insistence, in all areas of science, that a long table really is wider at the near end, because that's how a viewer sees it. But therein lies the flaw in Mach's thinking. An observer who walks around a table sees its appearance change. The conventional explanation is that there exists an independent object called a table, endowed with certain fixed properties, and its changing appearance is the result of looking at it from different perspectives. Reality, in other words, is the table itself, not the way the table happens to look from this or that angle. Similarly, in science, true reality is not what is seen directly but what is consistently inferred from a variety of observations. This is why Boltzmann and the atomists believed in the value of what they were doing. It might not be possible to see an atom directly, just as it is not possible to see the real table, free of the distortions of perspective, in any single view. But the table exists, and so do atoms.

Mach clung to his ideas fiercely throughout his life. Unlike Boltzmann, who had been an eager but docile schoolboy, Mach had resented instruction where it disagreed with his own notions. He liked to memorize the features of geographical maps or lists of historical events, but he struggled against conjugations and declensions in Greek and Latin, those being arbitrary and therefore unreal constructions to him. His obstreporousness may have lain behind the judgment of the Benedictine monks at one of his schools, who had found him "*sehr talentlos*." Mach's father, himself a schoolteacher, took his son away from the care of the monks and taught him at home for a while, and succeeded in getting the classical languages into his head. Later, Mach went to another public school, and once again had difficulty. Formal instruction again triggered rebellion in him. Like many children who know themselves to be intelligent but find themselves in difficulty at school, he attributed the success of others to a dubious kind of "school cleverness and slyness" that he himself lacked.

Nevertheless, he graduated from this school, and in 1855, at the age of 17, arrived in Vienna to study mathematics and physics at the university. Although he did well as an undergraduate, supporting himself by tutoring, he continued to complain. "Kaiser Franz had let the Austrian universities go to the dogs, and I didn't have enough money to go to a German university," he recalled much later. He remained "a stranger with respect to all of my professors, an outsider, someone they mistrusted and against whom they visibly tried to excite mistrust." Boltzmann, who arrived in Vienna a few years after Mach and studied mainly under the congenial Josef Stefan, remembered his undergraduate days as a period of encouragement and accomplishment compared to which his later academic experiences seemed less golden. Mach was there before Stefan arrived, and studied with the older, less distinguished, and considerably less modern physicist Andreas von Ettingshausen. Even so, the difference between Mach's recollections of the University of Vienna and Boltzmann's reflects their own personalities as much as any great change in the nature of the institution after Stefan took over. Boltzmann soaked up knowledge gleefully; Mach took it in bit by bit, critically, judging each idea against his own notions, undeveloped as they were at that stage, of what knowledge ought to consist of.

After graduating, Mach began to deliver lectures in elementary physics, earning some money so that he could also do experimental research in a variety of subjects. His lectures were popular, but printed booklets of his teaching failed to find many readers. At the same time–and quite unlike Boltzmann–he fell in with a varied group of writers, journalists, and critics who used to meet at the

Café Elefant in Vienna. Especially toward the end of the 19th century, the numerous coffeehouses of Vienna served as the foci for various earnest intellectual groups: the Freudians at one establishment, Mahler and his group at another, Trotsky presiding at a third. Coffee came into the city's life in a characteristically Viennese way, arriving by the agency of a would-be invader. The Ottoman Turks were, over the centuries, frequently at the gates of the city, and only after their final defeat at the end of a siege in 1683 was Vienna finally free of the Turkish threat. Still, the Viennese abstracted a number of choice elements of Turkish culture, coffee being first among them. The traditional story is that the fleeing Turks left behind bags of unroasted beans along with brewing equipment, and that a Polish spy working for the Austrians took charge of the mysterious items, since he was the only person who knew their purpose, and founded the city's first coffeehouse.

Mach owed his introduction to café society to his knowledge of the tone perception theories of Helmholtz. Overhearing a conversation led by a newspaper music critic, Mach was invited into the discussion and impressed the crowd with his ability to explain scientific aspects of the way music is heard. The classical education dinned into him by his father now yielded benefits, and he joined a diverse group of writers, musicians, and social philosophers. From the latter, in particular, he gained some understanding of broader issues that would in time come to inform his own thinking.

During his tenure in Prague, Mach developed from a versatile if not brilliant physicist into a more focused and purposeful philosopher. His scientific achievements included the invention of a laboratory demonstration of the Doppler effect, which became a staple of physics classes for years to come. He studied acoustics and fluid flow, made attempts at microphotography, worked at improving a medical device to measure blood flow and pulse rate, and began to think about human perception of shapes, colors, and sounds with the idea of coming up with physical explanations for physiological responses. He had even dabbled in atomic theory–not very successfully, as it turned out–in order to understand the flow of liquids through tubes. These wide-ranging activities attested to a fertile mind and an ingenious pair of hands, but they also betrayed an attitude that the job of a physicist was primarily experimental and observational. Even at this early stage, Mach distrusted theorizing and mathematizing except insofar as it might provide simple explanations for the quantities a physicist could measure in the lab.

From the beginning he had interests beyond the world of physics alone. In attempting to explore scientifically the means of human perception, Mach had come into contact with a long seam of philosophical thought. This in turn had led him to start thinking about the development of physics from a philosophical standpoint, and he began to attract a little—at first, a very little—attention as a commentator on scientific history and philosophy. In 1872, after he had been in Prague for five years, his book *The Conservation of Energy* appeared. In it, among other things, Mach made some attacks on the kinetic theory of heat and atomism in general. But the book was perhaps too philosophical for most scientists, too scientific for most philosophers, and generally too poorly argued for both camps. It quickly disappeared.

Mach wrote prolifically, both books and scientific articles, describing his experimental achievements, but for many years his writings slipped from the presses and sank out of sight with hardly a ripple. Indefatigable, he kept at it and gradually began to make an impression. As the years in Prague wore on, he turned away from experimental work and increasingly to writings of an expository and philosophical manner. After *The Conservation of Energy,* his 1883 monograph *The Science of Mechanics* was a much greater success and began to exert an influence on a younger generation of physicists.

Mach tried to draw a distinction between "mathematical physics," which was simply the elucidation of mathematical relationships between measurable physical quantities (and which was a good thing), and "theoretical physics," which connoted the attachment of deeper meaning, the attribution of some sort of reality, to mathematically defined quantities (and which was, by Mach's lights, a bad thing). At the end of *The Conservation of Energy,* Mach had declared that "the object of natural science is the connection of phenomena, but theories are like dry leaves which fall away when they have ceased to be the lungs of the tree of science." In much of his writing, Mach was at pains to scrutinize large areas of science, especially physics, and judge what were leaves and what was solid wood.

Even Newton did not come up to Mach's standards. He found that in Newtonian mechanics the concepts of "mass" and "force" are not defined independently from directly measurable quantities, but achieve definition only through the very laws they enter into. In other words, a mass can be defined from the force needed to move it, but forces are themselves defined according to their ability to move mass. This, Mach thought, was circular and unacceptable. He was half-right: Newton's laws do contain a degree of circularity, but the success of these laws is not that they are somehow self-evident, or can be derived from some more fundamental laws, but that they in effect define the subject they aim to describe. This is not a weakness but a strength.

To put it another way, any new scientific law has to rest on some kind of theoretical assumption. Newton showed not that mass and force could be independently defined in some unarguable way, but that the mass and force implied by his laws had universal meaning and applicability. This is indeed somewhat circular, but it has to be: Newton is erecting a theoretical edifice where none previously existed.

This was an aspect of scientific theory whose necessity and inevitability Mach could never grasp. He wanted all laws to rest only on definitions that had some obvious independent meaning; he would not, or could not, accept that science must devise qualities and characteristics whose usefulness can be proved only within the system that defines them. For scientific theories, the proof of the pudding is in the eating.

Mach's antipathy to theorizing and to the invocation of "metaphysical" and therefore unprovable notions led him to some extreme opinions. In *The Conservation of Energy* he remarks: "We say now that water consists of hydrogen and oxygen, but this hydrogen and oxygen are merely thoughts or names which, at the sight of water, we keep ready to describe phenomena which are not present but which will appear again whenever, as we say, we decompose water."

Mach is saying that it is acceptable to speak of hydrogen and oxygen when they are individually present, but that any suggestion that hydrogen and oxygen constitute water, or that water contains hydrogen and oxygen, is going beyond the bounds of reason. This is, clearly, a restrictive philosophy. It means that a scientist may only say "there was hydrogen and oxygen, but now there is water" or "there was water, but now there is hydrogen and oxygen." To hint that water is actually made of hydrogen and oxygen is to suggest a metaphysical connection that goes beyond the certifiable facts.

Mach's argument here is oddly reminiscent of a stage that very young children go through. Infants do not understand how a teddy bear can disappear behind a screen and then reappear on the other side; they tend to think that the bear has gone, do not know to look for it behind the screen, and regard the newly revealed bear as a quite new object. But very early, in an automatic developmental stage, infants realize that the bear is still there, but out of sight. They know it is behind the screen even though they can't see it. Evidently, the urge to attribute reality to objects we cannot see is something we all learn at an early age, and which is essential to our being able to navigate in the real world. But Mach, as with his difficulties in grasping the notion of perspective in art, seemed to put his strict conception of reason above simple common sense. He could not accept that hydrogen and oxygen continued to exist while they were hidden from him in the form of water.

Nevertheless, Mach's views began to win a following. To be fair, he preached an adherence to experimental facts and a caution against unfounded theoretical speculation, which were, and remain, important elements of scientific style. He emphasized the importance of simplicity, in the traditional sense of trying to find the simplest account of observed phenomena but also in a larger sense, that scientific explanation as a whole should constitute as simple and coherent a system as could be obtained. But his devotion to these reasonable causes was overzealous to the point of fanaticism, and in the end his philosophy, as far as it concerned the practice of science, amounted to a list of things that scientists ought not to do. Theorizing was prime among those sins.

The second half of the 19th century was the period when theoretical physics first began to establish itself as a subject in its own right. Max Planck recalled that when he was an undergraduate in Munich in the 1870s, he could not study theoretical physics because no such course was offered; he learned experimental physics and mathematics separately. But that was changing. Boltzmann went to Munich in 1890 specifically as a professor of theoretical physics, and he continued teaching in Vienna under that same banner. Kinetic theory and Maxwell's electromagnetic theory were the first great theoretical constructs, harmonizing large subjects by means of a physical model expressed in explicitly mathematical terms. Not everyone in physics saw this as a laudable development, however, and Mach's ideology became a flag for those who found the methods and ideas espoused by Clausius, Maxwell, and Boltzmann too abstract, too removed from the empirical world, to qualify as acceptable science.

Mach's return to Vienna, at the age of 57, became a triumph. His lectures, which ranged across history and philosophy, physics and psychology, drew large, rapt audiences. He was saying nothing that he hadn't been saying, and writing down in numerous dense volumes, for many years, but in Vienna he finally gained a wide following among intellectuals of all stripes. Mach had become increasingly enamored of an idea borrowed from the fledgling science of economics, according to which there was a marketplace of ideas that favored simplicity as a kind of efficiency: the most explanatory power for the least investment of hypothesis. This principle certainly should apply to physics, he argued, where it ruled against elaborate theorizing and in favor of mere description and observation, but it could also be extended into the arena of moral and ethical thought. Mach had a kind of Panglossian belief, coupled with an interpretation of Darwin's thinking on evolution, that from the general ferment of ideas and behaviors moral actions would emerge as those that benefited most of the people most of the time. In a Vienna fragmented by nationalistic and political divisions of growing vehemence, this may have seemed a soothing philosophy, suggesting that one should refrain from trying to explain what was going on in terms of mysterious and subterranean social forces, and trust instead that all would be well eventually. In the mid-1890s, Ernst Mach became a public intellectual of great repute, attracting some young physicists to his cause but also influencing poets and writers, musicians and artists.

All this must have been galling to Boltzmann. He had come to Vienna on the death of his mentor Josef Stefan, and very soon afterward, in July 1895, his old colleague Josef Loschmidt had died. "Did I return to Vienna as the gravedigger for those who had been so dear to me?" Boltzmann lamented in a memorial address. Loschmidt had sometimes been critical of kinetic theory–it was he who first voiced the reversibility objection clearly–but like the British physicists, he fundamentally believed in atoms and wanted above all to find out how they worked. Boltzmann never became close to the younger (and generally less talented) physicists who were in Vienna when he returned, many of whom came under Mach's influence. Boltzmann had hoped that in Vienna he would find serious-minded colleagues ready to debate atomism and kinetic theory; he found, instead, a university in the grip of a philosophy he could not understand, or if he did, thought it foolish.

ADDING TO Boltzmann's woes, the old reversibility objection came up again, in somewhat different form, and from a new corner. The great French mathematician Henri Poincaré had in 1893 proved a theorem showing that any closed mechanical system must, in the course of time, return to its starting point. This conclusion was relevant to a point that both Boltzmann and Maxwell had addressed but not resolved. In thinking about the way a set of atoms constantly moves from one possible dynamical state to another, they had assumed, without proof, a degree of randomness, so that a gas would visit all possible states in an essentially statistical manner. Poincaré's theorem proved that in at least one respect this randomness was not absolute. It stated with mathematical certainty that the system would at some point come back to its starting point and therefore begin repeating itself. Poincaré noted at the time that his result might prove troublesome for what he called "the English kinetic theories."

A couple of years later, a student of Max Planck's by the name of Ernst Zermelo made the attack specific. If, as Poincaré's theorem demanded, the atoms in a gas must sooner or later return to the exact configuration they began in, then Boltzmann's H-theorem could not always be obeyed. If the system evolved at first so that H decreased and entropy increased, then eventually it must go back the other way, with H increasing and entropy decreasing. Therefore, Zermelo said, the idea of kinetic theory that a gas of atoms would inevitably evolve toward equilibrium–maximum entropy–and stay there was simply wrong.

Although he had a powerful new theorem to back him, Zermelo was saying nothing that Loschmidt and then the English critics had not said before. Yes, Boltzmann agreed, a system might sometimes move in a way that would decrease entropy. Now Poincaré had proved that such things indeed must happen. But the question, as always, was how likely such events were. To say, even with mathematical certainty, that an event must happen is not to say that it will happen often, or even in a humanly imaginable period of time. Boltzmann took to the fray again, no doubt with some weariness. His published reply to Zermelo displays a mixture of sarcasm and petulance.

"Herr Zermelo's paper indeed shows that the relevant works of mine have not been understood; even so I am bound to be pleased by this paper as the first indication that these works have been given any attention at all in Germany," he declared in the introduction, and after going through all the technical matters concluded: "All objections raised against the mechanical view of nature are thus void of substance and based on error. He however who finds himself unable to overcome the difficulties that a clear exposition of gas-theory principles offers should in that case follow Herr Zermelo's advice, and resolve to give the matter up."

To make his point specific, Boltzmann estimated the approximate time that a simple system of about one trillion atoms, constituting one cubic centimeter of gas at room temperature, would take to return, as Poincaré said it must, to some precise dynamical state it had already passed through. The recurrence interval he came up with was a number of seconds containing trillions of digits—an unimaginable length of time. For comparison, he observed that if every star in the sky had the same number of planets as the Sun, and if every planet had the same number of people on it as the Earth, and if every one of those people lived a trillion years, the sum total of their combined lifespans would even so amount to a number of seconds with less than 50 digits. Poincaré's recurrence theorem might be mathematically unarguable, he concluded, but it was not of any practical concern.

He summarized Zermelo's objection in a more easily grasped way. From a strictly mathematical point of view, a set of 1,000 dice thrown enough times must eventually come up all ones. But such an outcome is fantastically unlikely. Zermelo, Boltzmann said, "is like a dice-player who . . . concludes that something is wrong with his dice because such an occurrence has not yet presented itself to him."

More subtly, the reconciliation between Poincaré's recurrence theorem and Boltzmann's H-theorem hangs on probabilities and timescales. Given a cosmic perspective, in which one is prepared to watch for limitless eons, Poincaré and Zermelo are correct: a system must eventually come back to its starting point. But on human timescales (and even on timescales of trillions and trillions of years), the likelihood of recurrence is negligible. In practical terms, therefore, the assumption that a gas explores all the available dynamical states in a random fashion may not be strictly true, but it's pragmatically close enough to true as makes no difference. Once again, Boltzmann's sense of the physics of the matter held true.

In both his *Nature* communication of 1895 and the response to Zermelo in 1896, Boltzmann enlarged on this point a little. In thinking of the universe as a whole, which was generally presumed at that time to be eternal, it might seem that everything would have to settle down into a perfectly uniform, perfectly stable equilibrium–clearly not the heterogeneous universe of stars and planets and empty space that astronomers were beginning to map out. The notion of an inexorable winding down of the universe into a featureless stasis had been pointed out by Clausius, who called it the "heat death." Boltzmann now suggested that even in such a state, there would be pockets that, strictly for reasons of chance, ran temporarily away from the general equilibrium and then fell back again. The corner of the universe currently occupied by humanity, he suggested, must be just such a place, where entropy happened to have hit a temporary low and was increasing again. Elsewhere there would be pockets of the universe where entropy was running down, and in such places, Boltzmann speculated, it might appear that time itself was running backward.

To Boltzmann, this may have seemed like a plausible speculation adding to the depth and interest of his kinetic theory. To his critics it seemed like a farfetched piece of backpedaling, indicative of the lengths Boltzmann had to resort to in order to defend his theory. Far from being a true and definitive result, the H-theorem, Boltzmann now seemed to be admitting, was true only some of the time, in certain special places in the universe. There remained, on the question of reversibility, no consensus. Few physicists had yet accustomed themselves to arguments involving matters of probability. Poincaré's theorem was perfectly true: a system must eventually come back to its starting point. Boltzmann accepted this, yet argued in his baffling way that somehow it wasn't important.

Zermelo made a further brief reply, notable in that he expressed considerable surprise that Boltzmann blithely admitted the second law of thermodynamics to be a probabilistic and not an absolute rule. Zermelo was Planck's student, after all, and even in the mid-1890s this notion shocked him. Still, he was not alone in his puzzlement. In England, William Thomson (who had become Lord Kelvin in 1892) had run up against the idea of theoretical predictions that amounted to calculations of probability, not assertions of certainty, and the seeming contradiction had brought him to a halt. In 1895, he wrote to Boltzmann (the two had corresponded occasionally for a few years) that "whenever other occupations allow me I return to it, but alas! I make absolutely no progress towards comfort of happiness in regard to it. This is very sad, as on it the whole of Thermodynamics hangs."

Zermelo's objection struck many physicists as acute, and Boltzmann's answers as evasive. Planck and Kelvin were respected and influential figures. Boltzmann began to feel once more unhappy, unappreciated, and alone. Mach's disciples began to refer to Boltzmann as "the last pillar" of atomism. Other young physicists recollected that "with few exceptions, the leaders in Germany and France were persuaded that the atomistic kinetic theory had already played out its role" and that the atomists "in those days fought somewhat on the defensive."

Maxwell was long dead, Clausius had died in 1888, and Stefan and Loschmidt were more recently gone. In Vienna in the 1890s, Boltzmann felt himself surrounded by his intellectual enemies, with no young physicists to support him. His letters to a journal editor in connection with the publication of Zermelo's objection and his rebuttal reveal how much Boltzmann felt himself isolated and vulnerable. "Now I come to a delicate point," he wrote. Observing that Planck was an adviser to the journal and Zermelo his student, he went on, "I believe I am entitled to demand: 1, that Herr Planck does not delay the appearance of my note; 2, that no word of it should be changed; 3, that no reply [from Zermelo] should appear in the same issue, but that they should answer later as they wish and are able to."

In the same letter Boltzmann portrays his beleaguered position: "Since I am, so it appears, now that Maxwell, Clausius, Helmholtz, etc., are dead, the last Epigone for the view that nature can be explained mechanically rather than energetically, I would say that in the interests of science I am duty-bound to take care that at least my voice does not go unheard." (In Greek mythology the Epigoni were seven warriors who sacked Thebes to avenge the deaths of their fathers in an earlier failed attack on the same city.) In another letter to the same editor Boltzmann muses, "Whether I will soon be alone in opposing the present direction of German science I cannot say."

Boltzmann's enemies were calling him "the last pillar" and he himself seemed to concur at times in that judgment. Even so, he strove to stay upright.

CHAPTER 8

American Innovations

New World, New Ideas

THE FIRST STEAMSHIP TO CROSS THE ATLANTIC arrived in New York on April 23, 1838, having left Cork, in Ireland, 19 days earlier. This vessel, the *Sirius,* was hardly the model of a transoceanic liner. A typical packet boat of the sort that had been crossing the English Channel routinely for some decades, it had been hastily modified to make the transatlantic voyage in order to beat the new liner *Great Western,* built by the celebrated English engineer Isambard Kingdom Brunel. To complete its trip to New York, *Sirius* had to burn some of its cargo during the last couple of days. *Great Western,* delayed by various mishaps, arrived a few days later, but after a speedier 15-day journey and with fuel to spare. It had been thought by many engineers that no ship could carry enough fuel to make it all the way to the New World, a point that *Sirius*'s adventure reinforced. Brunel, however, had made the scientific observation that resistance to a ship's motion through the sea increases in approximate proportion to its surface area, while its coal-carrying capacity depends on the ship's volume. A larger ship, therefore, could take on proportionately more fuel, and complete the voyage comfortably.

Even so, the success of *Great Western* hardly made transatlantic travel routine. This ship used sea water in its steam engines, which as a result clogged and corroded quickly. The pistons had to be dismantled and cleaned regularly, which meant they could only be of an inefficient, low-pressure design. In 1856 came the innovation of a new condenser for such engines, which allowed them to recycle a limited quantity of fresh water. The engines could be sealed tight and operated at higher pressure. The new marine steam engine, as much as any innovation, began to make ocean-crossing steamships commonplace, and by the 1860s, travel from Europe to America was becoming more frequent and relatively cheap.

The man who invented the new condenser was the Scottish physicist William Thomson, later Lord Kelvin. If physicists in Vienna were inclined to debate philosophy in their spare time, those in Britain and Germany were more likely to tinker with steam engines or the telegraph. Boltzmann was something of an exception. Before his eyesight became altogether too poor for laboratory work, he had been a talented experimenter, and something of a domestic tinkerer too. He had once rigged up an electric motor to run his wife's sewing machine. In 1879, he presented a short note to the Viennese Academy of Science on some aspects of electrical theory relevant to telephone communications (Alexander Graham Bell had taken out his first telephone patent just three years earlier). Late in life, stimulated by the interest his surviving son, Arthur, showed in ballooning, Boltzmann became fascinated by the possibility of powered flight, to the extent of arranging some small experiments on suitable engines and giving lectures on his observations. He joined the newly founded Electrical Engineering Society of Vienna and even served as its president for a time. On one occasion he organized a little party at his house in Vienna to demonstrate a new light bulb invented by his friend the chemist Walter Nernst. He and his wife ordered 50 liters of beer, a cold buffet, and sent out 55 invitations. Seven of his colleagues showed up to see the new technological marvel.

Boltzmann was also quicker than most of his Viennese contemporaries to take advantage of the burgeoning steamship industry. Always a great traveler, he had before the end of his life been to America three times. His first Atlantic crossing came in 1899, when he was 55 years old. He had been awarded an honorary doctorate by Clark University in Worcester, Massachusetts, then celebrating its centenary, and he went there to pick up his degree and deliver a series of lectures on mechanics. The trip would be, he observed, a break from "the monotonous working life of Vienna," but when he reached Worcester he professed to find it "rather boring." After receiving the invitation, he had written to G. Stanley Hall, the president of Clark University, to explain that his health had become much poorer after he left Munich for Vienna and that because of his nervous state, it was essential for his wife to come with him–so that he would need a little more financial support for the expedition.

In June 1899, he and Henriette made the lengthy voyage to New York, which they found impressive but dangerous because of the speeding trams. From there they traveled to Boston ("the dust is terrible," Henriette wrote on a postcard) and thence to Worcester. They also took trains up and down the East Coast, going as far north as Buffalo and Montreal (taking in the Niagara Falls) and south to Baltimore and Washington. The travels were a combination of tourism and academic visiting. There were a few American physicists, mostly experimentalists, whose reputations were known in Europe.

Boltzmann did not, however, meet the single American scientist whose endeavors were most closely aligned with his own. In New Haven, Connecticut, resided Josiah Willard Gibbs, who like his Austrian counterpart was a pioneer of statistical thinking in physics. Boltzmann was certainly aware of Gibbs and his work. But Gibbs was something of a shadowy figure, and there were reasons Boltzmann may have thought him more an intellectual adversary than an ally.

Gibbs was an enigmatic and little-appreciated figure even among his own countrymen, in part because of the seeming abstruseness of his work, in part because of his personality. Born in 1839 in

New Haven to a professor of sacred languages in the Yale School of Divinity, he spent his entire life from the age of about seven in the house his father had built, close to the university. In death, he progressed no farther than a cemetery two blocks away. After a three-year excursion to France and Germany to round off his education, Gibbs never left America again, and indeed rarely left New Haven. Five years older than Boltzmann, he had spent a year in Berlin and then a year in Heidelberg shortly before Boltzmann's first visits to those places, but he occupied his time there in studious invisibility. The professors whose classes he attended recalled no anecdotes about him, and he came under the influence of no particular teacher. He never married, and on returning to New Haven settled in his now-deceased father's house with an older sister, who also never married.

But he was not a hermit. Another sister had married, and the nieces and nephews were fond of their Uncle Willie, as he was of them. He enjoyed walking and riding vacations in the New England countryside and would drive the horse and cart to take the children on picnics and adventures. Students found him a patient and encouraging teacher, thoughtful and occasionally amusing. He had little time for scientific societies and organizations, but he corresponded warmly with a number of individual scientists, including many in Europe. His work was known and respected by those who knew about such things, but he expended no extra effort to explain himself beyond the confines of his scientific papers, which even his admirers found clear but terse to the point of indigestibility. He was a profoundly self-contained man, in both his work and his life, and seems to have been perfectly happy in that condition. "Effusiveness was foreign to his nature," one colleague recalled, rather understating the matter.

Gibbs was, in a sense, self-taught. As an undergraduate, he had learned mathematics and classics, with only a rudimentary selection of science, because that was all that Yale had to offer at the time. Physics was covered in a single course, completed in one year under the rubric "natural philosophy" and encompassing a little chemistry, astronomy, mineralogy, and so on. After his degree he enrolled as a student of engineering, which included the novelty of laboratory work–something his previous science studies had not covered. His first small thesis was on the design of gears. Shortly afterward, he obtained a patent for an improved design for a railway car brake. In 1863, he was awarded a Ph.D., only the second scientific doctorate to be given in the United States and the third overall.

In 1866, with his two sisters, he went to Europe, visiting France and then Germany. They supported themselves modestly but adequately with family money. Gibbs worked hard to acquire a more sophisticated knowledge of mathematics and physics, which no American university could then provide. All that survives is a selection of his characteristically succinct lecture notes, with no real indication of whether any subject particularly excited or repelled him. On his return, he minded his own business for a couple of years, but then Yale decided it needed to modernize its teaching, particularly in science. Gibbs was taken on as a professor of mathematical physics, for no pay. At that time he had published nothing besides a few small ventures in engineering design, but he became known as an effective and diligent teacher. He seemed destined for a quiet and uneventful life, in keeping with his character.

But just a few years later, Gibbs burst out of his anonymity with three redoubtable scientific works. The first two shorter items appeared in the *Transactions of the Connecticut Academy of Sciences* in 1873. The third and much longer work was published in the same journal in two sections, in 1876 and 1878. With these publications, Gibbs transformed the state of classical thermodynamics.

The first two papers consisted, in essence, of a variety of novel graphs and diagrams. Scientists by this time had accustomed themselves to describing the states or phases of materials–gases, liquids, and solids–in terms of pressure, volume, temperature, energy, entropy, and so on. In his characteristically thorough and systematic way, Gibbs went through all possible combinations of these properties and showed that the resulting graphs (pressure versus temperature for a fixed volume, for example) could answer all kinds of questions. Some of the graphs turned out not to be especially useful, but one in particular–a graph of entropy versus volume–proved ideal for studying the conditions under which a gas would transform into a liquid, or a liquid into a solid. The second paper went on to show the utility of three-dimensional graphs whose axes represented three thermodynamics properties. Specifically, he described a graph plotting entropy, energy, and volume against each other. For any substance, this three-dimensional space divides into regions corresponding to the gaseous, liquid, and solid states; the boundaries between those regions form surfaces that, Gibbs showed, contain further information. For example, the gradient of the surface separating the liquid and the solid regions is related to the temperature and pressure at which solid and liquid transform into each other.

This may sound a little simple-minded, as if a great scientific discovery emerged just by drawing a lot of graphs and pictures, but it took great insight, in the 1870s, to see how such diagrams combined a wealth of formerly disparate notions about stability, equilibrium, heats of evaporation and condensation, and a host of other physical attributes of materials. Gibbs' achievement was to see how these deceptively straightforward devices could be exploited to yield a wealth of information. Maxwell, for one, was delighted. In 1875, he delivered in a lecture to the Chemical Society in London to alert British scientists to "a most important American contribution to . . . thermodynamics." By the use of Gibbs' methods, he went on, "problems which had long resisted the efforts of myself and others may be solved at once." So enchanted indeed was Maxwell that out of plaster of Paris he constructed a three-dimensional "thermodynamic surface" for water, according to Gibbs' design, and sent it off to Yale, where it resided proudly on Gibbs' bookshelf.

But the third installment of Gibbs' work was by far the greatest. In two papers amounting to more than 300 pages, Gibbs tackled as completely and comprehensively as he could the question of thermodynamic stability. Previously Gibbs, like everyone else, had been mainly thinking about the thermodynamics of single substances, and how the transitions from solid to liquid to gas depended on physical conditions. The element of genius in his 1878 work was to perceive that essentially the same analysis could be applied to mixtures of more than one substance, and to changes other than transformations of physical state. A simple example would be moisture in air: Under what conditions does it remain suspended, and when does it condense out into droplets? A more complex case would be a solution containing several chemicals that react with each other in a variety of ways: When is one reaction preferred over another? When will a solid product drop out of the solution, and when will it dissolve back again? These are all, Gibbs realized, questions that can be answered thermodynamically.

Moreover, he recognized that it is only a matter of practical complication to add as many substances and as many phases and as many interactions as one cared to. Most notably, he could include chemical changes, in which components of a system reacted to form a new component, perhaps giving out energy in the process. He could include decompositions, in which one component broke apart into two others when the temperature reached a certain value. All such changes came down to questions of stability: in some circumstances, it was preferable for two components to remain separate; in other circumstances they might react or dissolve. Any mixture of interacting components could be treated in the same way. It was, Gibbs demonstrated, merely a matter of keeping track of all the possible states each component could be in, and all the interactions they could participate in.

So complex was this analysis that the drawing of graphs was inadequate to the task. Nevertheless, the principle was the same, and Gibbs patiently set out a comprehensive system of algebraic equations that embodied any physical system he wanted to understand. The key to his technique was both simple and versatile. Imagine, Gibbs instructed, what would happen if some tiny element of a large and complex system were to change in some way.

A little volume of gas might become a liquid; a dissolved component might separate out; a chemical compound might break up into its constituents. Any such change has thermodynamic consequences, producing corresponding changes in energy, pressure, temperature, entropy, and so on.

Taking all this into account, Gibbs asked, when such a change occurs, does the system as a whole turn into an energetically more favored or less favored state? If a small change leads to an overall lowering of energy, then the system as a whole will spontaneously undergo the change and transform into a different state. If, on the other hand, the imagined change costs energy, then the system will stay where it is.

By this method, Gibbs established a universal and completely general way to analyze the stability of any system. The same technique made it possible to understand how the system would react when external conditions were changed. It might be heated up or cooled down, expanded or compressed; what would all the internal components do? His way of answering that question was to examine (algebraically, that is) all the possible consequences of change to small units of the system as a whole, and to sort out which were energetically favorable.

The end result of this great effort was that Gibbs showed how to calculate questions of stability, mixing, and equilibrium, for any mix of ingredients, defined strictly by their physical properties. His method relied only on the thermodynamic properties of materials, making no assumptions of any kind about their fundamental composition, atomic or otherwise. This indeed was both its greatest virtue and, for some, the greatest barrier to understanding.Gibbs' strategy was novel. His aim was to construct a perfectly logical and rational system into which the interested physicist, chemist, or engineer could plug in whatever details were appropriate to the question at hand. His method was as suitable for understanding when raindrops would condense out of moist air as it was for calculating how impurities of carbon would dissolve in molten iron.

So versatile was Gibbs' technique that to many chemists and

engineers–the audience for whom it would prove most useful–it seemed fabulously abstract. His style of exposition, which put a maximum of meaning in a minimum of words, didn't make his work any easier to grasp. (Some years later the British physicist Lord Rayleigh wrote to Gibbs asking if he might not be able to produce an augmented and amplified version of some of his works, easier for less expert readers to comprehend. Gibbs replied that "I myself had come to the conclusion that the fault was that it was too *long*.")

Nevertheless, Maxwell got the point immediately, and wrote a short notice for the *Proceedings of the Cambridge Philosophical Society* expressly to bring Gibbs' ideas to the attention of British scientists. In Germany, his ideas struck Wilhelm Ostwald as a revelation. Expressing the laws of chemical and other transformations in physical terms, with matters of stability tied directly to considerations of the overall energy and entropy of the system, was exactly his goal, the essence of the subject that came to be called physical chemistry. Ostwald had been going at the problem piecemeal, and now an unknown American had written an encyclopedia on the subject all at once.

Although the *Transactions of the Connecticut Academy of Sciences* was not widely in evidence among the university libraries of Europe, Gibbs took care to send copies of his works to some hundred or so scientists he presumed would understand and appreciate them. Maxwell and Clausius, Helmholtz and Ostwald, were on the list; the name of Boltzmann, who was not yet well known in the early 1870s, was added for the third of the three papers. Ostwald translated the papers into German and arranged for their publication. In his autobiography he recalled: "This work had the greatest influence on my development. For, although he does not especially emphasize it, Gibbs deals almost exclusively with energy and its factors and holds himself free from all kinetic hypotheses. Because of this, his results possess a certainty and a lasting quality of the highest degree humanly attainable."

Ostwald's evaluation was accurate but shaded by his own prejudices. It was true that Gibbs did not indulge in any "kinetic hypotheses"–that is, Gibbs did not make any specific assumptions about the nature of atoms or their motions and interactions. But the point was not that he was against such ideas, rather that he didn't need them for his own purposes. Ostwald takes this omission, along with Gibbs' emphasis on the use of standard thermodynamic properties, as covert support for his own philosophy of energeticism. Gibbs' ideas seemed to follow Mach's prescription of basing theoretical arguments on tangible physical characteristics, not speculative abstractions, and it seemed to adhere to Ostwald's preference for making transactions of energy the fundamental principle for deciding stability or instability of any system.

But Gibbs had no interest in Mach's philosophy and showed no particular liking for energeticism. He dealt straightforwardly in well-defined thermodynamic properties, and eschewed any speculation as to the true nature of the world, or the correct philosophy to employ. His agnosticism was the strength of his work, but it allowed believers in any number of philosophical camps to believe he was secretly one of them.

Perhaps because of Ostwald's enthusiasm for Gibbs' work, Boltzmann's attitude to the new ideas coming from Yale was ambivalent. Where Ostwald took Gibbs to be a closet energeticist, Boltzmann imagined instead that he was using atomic ideas but then concealing the fact. "In justifying his theorems, Gibbs must surely have used molecular ideas, even if he nowhere introduced molecules into the calculation," he wrote on one occasion. Elsewhere, Boltzmann refers to a Gibbs theorem "which he had discovered by a different method though still presupposing certain basic conceptions of molecular theory." That both Ostwald and Boltzmann could see Gibbs as a potential ally surely testifies to his philosophical neutrality.

In one important respect, however, Gibbs was not only sympathetic to Boltzmann's views but ahead of them. Like Maxwell, he perceived the fundamental importance of probability and statistics in this kind of physics before Boltzmann somewhat reluctantly tackled the issue. He arrived at this insight, typically, in his own way. Gibbs' general analysis relied on considering a large system as being composed of numerous small units, each with its own thermodynamic properties, and then deducing the behavior of the whole as a consequence of what all the component parts were doing. He therefore grasped early on the fundamentally statistical nature of such systems. The question came up clearly in a curious observation that has become known as Gibbs' paradox.

Imagine a chamber divided into two halves, separated by a removable partition. Suppose first that the two halves are filled with different gases at the same temperature and pressure, and that the partition is lifted. The gases would mix and, as Gibbs demonstrated with his new style of reasoning, there would be an increase in entropy precisely because they had mixed.

Now, Gibbs went on, imagine the same thing except with two identical volumes of the same gas in the two sides of the chamber. Lifting the partition and allowing the gases to mix could not, in this case, produce an increase in entropy because, for practical purposes, nothing has happened. The two volumes of gas will certainly mix, but since they are the same gas, no perceptible physical change follows. The thermodynamicist cares only about overall properties; it doesn't matter that elements of one gas that were formerly on one side of the partition have moved to the other, or vice versa. The details of how all the microscopic elements of the gas get themselves mixed up with each other are of no consequence. The entropy is the same regardless.

This difference–entropy goes up when different gases mix, but not when identical gases mix–has become known as Gibbs' paradox largely because of the consternation it still causes to the average (and even above average) physics student. There's no sign Gibbs himself found anything paradoxical in it. Rather, he saw it as a demonstration that thermodynamic properties cannot depend on microscopic details of exactly which atom is where. But he used it to make a further characteristically astute observation (and one, incidentally, in which he specifically talks of "molecules" of gas).

When identical gases mix, it doesn't matter where all the molecules go after the partition is lifted. All motions are equal; they intermix at random and the entropy stays the same. But now imagine putting labels on all these molecules so that they correspond to two distinct gases. Their movement is unaffected, but now the entropy depends on how the labels distribute themselves. There must be some sets of molecular motion, Gibbs said, which put one kind of molecule mostly on one side of the chamber and the second kind on the other. The argument concerning identical gases says that that must be a physically allowable possibility, but applied to the case of different gases, the same possibility amounts to a spontaneous separation of the gas into distinct halves, which would correspond to violation of the second law. "In other words," as Gibbs put it, "the impossibility of an uncompensated decrease of entropy seems to be reduced to improbability." By "uncompensated decrease" Gibbs means a reduction in entropy without there being a concomitant increase somewhere else; this is what the classical second law is supposed to forbid.

These words come from the first installment of Gibbs' long paper, which appeared in 1876. That was the year after Boltzmann had been faced with Loschmidt's objection to the H-theorem, which forced him to recognize the very possibility that Gibbs has hit on, and the year before he published his famous formula, $S = k \log W$, connecting entropy with statistics–a result that Gibbs' own reasoning anticipates.

Uninfluenced by either the Vienna or the Cambridge way of thinking, Gibbs developed a new way to think about probability in thermodynamics, a way that depended not at all on arguable assumptions about the nature or existence of atoms.

Boltzmann understood his microscopic analysis to refer specifically to systems of atoms, and he understood the changes in them to be specifically changes in the distribution of atoms. Gibbs, by contrast, thought of a system in terms of microscopic components and changes, but those components were defined entirely by their thermodynamic qualities. As in the preceding example, he used the word molecule from time to time, but he held no fixed notion about what these molecules might be. It didn't matter. This gave his analysis a greater power than Boltzmann's, since he could imagine any thermodynamic property he liked, not just those for which an atomic model was at hand.

But contrary to what Ostwald wanted to believe, Gibbs was not in any strident sense an anti-atomist. He saw no reason to take one side or another. It was a practical matter: "we avoid the gravest difficulties when, giving up the attempt to frame hypotheses concerning the constitution of material bodies, we pursue statistical inquiries as a branch of rational mechanics." As Ostwald observed, Gibbs' methods derive their strength and universality from their independence of any assumptions about the nature of matter, but at the same time there is nothing in Gibbs' writings to suggest that he had any fundamental antipathy to atomism, or any fondness for the strictures of energeticism.

Gibbs' published treatise of 1876 and 1878 was, for all its power, both apparent and implicit, in some ways an obscure work. Perhaps not even the author grasped all its implications, and it was some 20 years before Gibbs himself returned to the subject in order to set out in a book a fuller and more systematic account of his methods.

Nonetheless, in the late 1870s and early 1880s, as Gibbs' work became known in Europe, it could have been embraced by Boltzmann as a powerful complement to his own ideas. Gibbs had no need for any atomic hypothesis, but at the same time, the fruits of the atomic approach–Boltzmann's statistical definition of entropy, for example–could have been allied to Gibbs' methods and used to extend the versatility and range of atomic theory. Boltzmann, in other words, might have used Gibbs' analysis as a means to systematically explore the wider implications of his own vision of the nature of heat.

Boltzmann was certainly aware of what Gibbs had done, but he seems not to have fully appreciated its scope. For one thing, Gibbs' long papers concerned themselves in minute details with complex systems of mixed substances of a general nature, gaseous or otherwise, interacting chemically as well as physically, and this may have seemed to Boltzmann far removed from his own more specific concern of trying to understand the behavior of individual gases from the motion of their atoms. Where Boltzmann mentions Gibbs, he seems mostly at pains to suggest that Gibbs is really using atomic models and reasoning, if in a somewhat surreptitious way. In the second volume of his *Lectures on Gas Theory,* for example, Boltzmann says "in many places it seems evident that Gibbs has these molecular-theoretic concepts continuously in mind, even if he does not make use of the equations of molecular mechanics." More strikingly, Boltzmann's introduction to this volume uses as a motto Gibbs' phrase about the "uncompensated decrease" of entropy being an improbability rather than an impossibility. This again is an instance of Gibbs seeming to be in Boltzmann's corner, since the energeticists and assorted other followers of Mach and Ostwald took the second law to be absolute.

Mostly, though, it was Ostwald who seized on Gibbs as an ally, arguing (despite Gibbs' own occasional use of the word molecule and his express desire to avoid fundamental assumptions as to the nature of matter) that he was really engaged on a project to free thermodynamics of any "metaphysical" assumptions. Gibbs himself, comfortable in New Haven, is not known to have expressed any views on his being co-opted in this way. Most likely he would have been quietly amused at being misinterpreted by both sides, and would have kept his amusement to himself.

DESPITE A NUMBER of chances, Gibbs and Boltzmann never met. After his one trip to Europe as a young man, before Boltzmann's name was well known, Gibbs remained on his side of the Atlantic. In the late 19th century, physicists in Europe were not in the habit of flitting over to the United States and would have seen little reason to go there even had travel been less inconvenient. Gibbs was invited to meetings of the British Association in 1887 and 1893, but didn't go. He was not invited, apparently, to the 1894 B.A. meeting in which kinetic theory was a prime topic and in which Boltzmann so enthusiastically took part. This omission may indicate a general belief that Gibbs' work was not seen as directly relevant to atomism and kinetic theory. Or it may be that British scientists had figured out that Gibbs was not likely to come anyway.

Boltzmann himself invited Gibbs to a scientific meeting in Nuremberg in 1892, but the American once again declined, and when Boltzmann went to the United States in 1899, he made no effort to seek out Gibbs. A few years later, in 1901, Yale celebrated its bicentennial, and an invitation to the festivities was sent to Boltzmann. But by then Boltzmann's health was deteriorating, and he didn't go. In any case, it's far from obvious that a meeting between the two would have produced any enlightenment. During the 1890s, Gibbs was mainly occupied with teaching and with a variety of less consequential researches than his earlier endeavors in thermodynamics; he was not to return to them until the early 1900s, when he was persuaded to assemble his arguments and thoughts in an influential book called simply *Elementary Principles in Statistical Mechanics*. Boltzmann meanwhile was embroiled in philosophical debates over what theories of physics ought to do, and his published scientific research was less momentous.

Then too there was the matter of personality. Gibbs would no doubt have got on famously with Maxwell; both were dry, ironic characters, capable of a sharp wit, affable in private but undemonstrative in general. And both, for all their prowess and insight in physics, had a capacity for not taking things too seriously. Maxwell had contributed to many areas of physics, and he looked upon each one of them as profound in its own right but not necessarily indicative of some vast underlying system to physics in general. Gibbs too, in his careful and exhaustive way, had raised thermodynamics to a new level of sophistication and logical order, but he saw it as a self-contained subject, enormously powerful within its own well-stated confines but indifferent to the underlying nature of matter. Gibbs and Maxwell shared a kind of wry detachment from even their own greatest creations, and a sense of irony is a sure antidote to any temptation toward grand philosophical systems. Indeed, Maxwell had written on one occasion to his friend Tait that "I have read some metaphysics of various kinds and find it more or less ignorant discussion of mathematical and physical principles, jumbled with a little physiology of the senses. The value of the metaphysics is equal to the mathematical and physical knowledge of the author divided by his confidence in reasoning from the names of things." The second sentence is a characteristically sharp Maxwellian gibe: it means literally that the greater a philosopher's confidence in his thinking, the smaller the value of the resulting cogitation.

Boltzmann, however, returning to Vienna from Munich, found himself beset on all sides by philosophers and energeticists who disputed the very essence of kinetic theory. He lacked, furthermore, a capacity for detached amusement that would have enabled him to ignore their harangues and complaints and to see that their arguments were bound to fail in the end. He took it all very seriously. He had devoted the bulk of his intellectual life to a single question–how does the behavior of atoms explain why hot things always cool down?–and in his single-mindedness he was often humorless, obstinate, and fierce. Could Gibbs and Boltzmann have even conducted a conversation? Would the laconic New Englander have reacted to the voluble and irrepressible Austrian simply by lapsing into silence, nodding occasionally to show he was listening, and hoping for the hour to be up when he could make an excuse and leave? It's tempting to suppose that an earnest conversation between Gibbs and Boltzmann in Nuremberg in 1892 might have sorted out some puzzles that troubled Boltzmann throughout the following decade, or bolstered his confidence that the philosophical attacks were not worth the time he spent on them. But it's equally easy to imagine such an encounter as a spectacular mismeeting of minds, leaving the austere Gibbs quietly stunned and the manic Boltzmann perplexed and frustrated.

After returning to Europe from his first trip to the United States and before taking up the reins of teaching once again, Boltzmann took a short break in Abbazia, a resort now known as

Opatija on the Adriatic coast of Croatia but at that time an Austrian possession. He had traveled a good deal in the previous year, visiting Göttingen, London, and the Netherlands for a variety of reasons, including official duties on behalf of the Viennese Academy of Sciences. His health was increasingly troublesome. His worsening eyesight forced him to pay an assistant to read scientific papers to him, and he was incapable of any further work in the lab. Over the years he had gone from plump to fat to corpulent, and he suffered sporadically from bladder problems, asthma, and what he called catarrh, the latter being a sort of catchall term for a variety of ailments ("stomach catarrh" was a recurring nuisance, something that might today be called, with equal inaccuracy, stomach flu). Physical ill-health upset him mentally, and in the midst of a Vienna in which he had more intellectual enemies and detractors than supporters, he suffered from episodes when he felt defeated and forgotten. His triumph at the debate in Lübeck (if he even counted it as a triumph) was in the past, and the continuing attacks on the H-theorem, no matter how many times he rebutted what was essentially the same question in different guises, began to dent his confidence. In 1898, he had written to Felix Klein, his assistant at the Lübeck debate, "Just when I received your dear letter I had another neurasthenic attack, as I so often do in Vienna, although I was spared them altogether in Munich. With it came the fear that the whole H-curve was nonsense."

In the introduction to the second volume of his great monograph *Lectures on Gas Theory*, published in 1898, he publicly expressed his assessment of his subject and his own position. "I am conscious of being only an individual struggling weakly against the stream of time. But it still remains in my power to contribute in such a way that when the theory of gases is again revived, not too much will have to be rediscovered."

And in September of the following year, Boltzmann delivered a historical and philosophical survey entitled "On the Development of the Methods of Theoretical Physics in Recent Times," in the course of which he observed, "I feel like a monument of ancient scientific memories. . . . I regard as my life's task to help ensure . . . that the great portion of valuable and permanently usable material [in my work] need not be rediscovered one day."

By the turn of the century, Boltzmann had been back in Vienna, his home town, for six years, and he was lecturing and writing in the wounded tones of a man composing an obituary notice for himself that would only be appreciated by an unborn generation.

CHAPTER 9

The Shock of the New

The Arrival of the Atomic Century

IN NOVEMBER 1895, A GERMAN PHYSICIST named Wilhelm Röntgen discovered a new and penetrating kind of radiation, with the power to pass through a person's flesh and yield an image of the bones concealed within. They were dubbed X rays, in the old scientific tradition of X standing for the unknown, and as news of the discovery raced around the world, physicists everywhere dropped what they had been doing and built X-ray generators.

In Paris, early in 1896, Henri Becquerel attended a lecture describing the new radiation, and wondered if there was any connection between X rays and the phenomenon of fluorescence, in which some materials continue to glow after they have been exposed to ultraviolet light. He happened to have a selection of fluorescent minerals and crystals in his laboratory and began some new experiments. He accidentally left one such rock in a laboratory drawer for several days, where it sat on top of some photographic plates wrapped in black paper. When he came to use the plates later, he found they had been ruined, fogged by some unknown emanation. The piece of rock was rich in uranium compounds. Becquerel had discovered radioactivity, and before long, Marie Curie had embarked on her painstaking and, as it turned out, dangerous task of sifting through tons of uranium-bearing minerals to tease out and identify the new radioactive elements.

The following year, in Cambridge, J. J. Thomson was investigating another kind of radiation, called cathode rays, which physicists had learned to create inside electric discharge tubes. He demonstrated that the rays were in fact a stream of particles, electrically charged and with a mass much less than the estimated mass of atoms. They became known as electrons.

X rays, radioactivity, subatomic particles. In the last five years of the 19th century, physics was turned upside down by a string of unexpected discoveries. X rays were eventually identified as an unexpected form of Maxwell's electromagnetic radiation; the other discoveries were beyond the confines of the physical universe as it was then understood. They were new forms of energy and matter, and they became the foundation for nearly all of the new physics that emerged in the 20th century.

In Vienna, meanwhile, Ludwig Boltzmann was teaching classical physics to students he found complacent and unpromising, and was mired in a sluggish philosophical debate, which, after an initial burst of excitement, gave him little pleasure and a great deal of frustration. Boltzmann took virtually no part in the explorations of this new world of physics. In 1895, the year after his return to Vienna, the most momentous event in his life had been not the abrupt discovery of X rays, but the arrival of Ernst Mach as a professor of philosophy. Boltzmann found few worthy students and colleagues in Vienna, and those already there seemed to show more interest in mulling over Mach's wide-ranging philosophical critique of physics than in actually doing any worthwhile physics themselves. Inevitably, Boltzmann found himself drawn into the debate. He was not a philosopher, either by education or by inclination, but like anyone who had gone to school in a German-speaking part of the world, he had absorbed a little smattering of Kant and at least knew some of the words that would gain him entrance into the world of Mach and his followers.

Over the years, Boltzmann had made an effort to follow Mach's views and even to understand his perspective, but he found the philosophy so alien to his own way of thinking that the task of rebutting it seemed both futile and infinite. It was obvious, to Boltzmann, that a version of science based on Mach's dictates against theorizing would be anemic in the extreme. There had to be more to physics than merely listing observations and attempting to find simple mathematical connections between them. In Mach's world, there was to be no such thing as "explaining" in the way scientists had always understood it. Mach even went so far as to argue that the traditional notion of cause and effect–that kicking a rock makes it move, that heating a gas makes it expand–was presumptuous and therefore to be denied scientific status.

Mach had said, for example, that "the law of causality is sufficiently characterized by saying that it is the presupposition of the mutual dependence of phenomena. Certain idle questions, for example, whether the cause precedes or is simultaneous with the effect, then vanish by themselves." He meant, in other words, that it might well be the case that when a volume of gas was heated, it would be seen to expand in volume, but to presume from this that the heating causes the expansion is quite unwarranted.

It's hard to make any useful theory in physics when even the most elementary notion that one thing causes another is ruled out of order. That was exactly Mach's point. He wanted to expunge theorizing from physics altogether and to leave only the bare cataloguing of correlations, a listing of which phenomena reliably happen in conjunction with others. It was permissible to make quantitative relationships. A certain amount of heat applied to a certain volume of gas will consistently be seen in association with a certain degree of expansion. But no, said Mach; one must not suggest that the heat causes the expansion.

Taken to such extremes–and there's no doubt that this is what Mach had in mind–this so-called anti-philosophy seems absurd. So indeed it seemed to Boltzmann, but arguing against it, in terms that philosophers such as Mach would understand or even allow, proved problematic.

The essential difficulty is that most physicists have no interest in philosophy and see no need for it. They may not know much about it, but they know what they don't like. Science demands an element of creativity, and an element of faith. The creativity comes in thinking up hypotheses and theories that no one has ever thought of before. The faith comes in thinking that these hypotheses, when shown to be useful or successful in some way, bear a relation to what is loosely called reality.

Lucretius, when he wrote *On the Nature of Things,* believed that atoms genuinely existed, even though they were undetectable to human senses. Their actions, he hoped, would reveal them. Boltzmann and the other atomists, at the end of the 19th century, believed the same thing. Philosophically, they had made no progress, but scientifically, they had made atoms far more real–or better to say, more credible. By the 1890s, physicists could calculate properties and behaviors of tangible physical objects by attributing to the atoms specific characteristics and assuming that they obeyed Newton's laws of mechanics. They had replaced Lucretius's vague and picturesque yearnings with a quantitative mathematical theory, out of which testable laws emerged. This, in Boltzmann's view, was the essence of turning an idea into a theory, of making it a piece of science and not a work of the imagination.

Mach understood all this, but refused to accept it. No matter how successful atomic theory might be–and even at the end of the 19th century its successes remained arguable–it was impossible to prove that atoms were real. Atomism might, Mach allowed, constitute an acceptable model, a set of mathematical relations that yielded correct answers, but nowhere in any of that can there be any objective proof that atoms existed. The fact that the theory worked was not enough to prove that the assumptions on which the theory rested were true. That was a circumstantial case only.

This is the heart of the dispute between Mach and Boltzmann. It remains to this day unresolved, because the two sides are not debating the specifics of any theory, but applying different standards to what a theory should be. When Mach stood up in the

Viennese Academy of Sciences and objected to Boltzmann's arguments with the words, "I don't believe that atoms exist!" he was saying, more precisely, that he would not allow himself to believe that atoms existed unless Boltzmann or someone else could prove beyond doubt that they must exist. Boltzmann knew perfectly well that he could provide no proof of the sort that Mach demanded. His belief in the existence of atoms–and belief is the right word–derived from the uncommon good sense that kinetic theory made. Assume atoms exist, assume that they follow the usual mechanical laws, and all else follows. To Boltzmann that was enough, indeed all that any scientist could reasonably demand.

Boltzmann was further handicapped in that he was working within (or rather, against) a German tradition of philosophy. Kant was the last in a long line of philosophers, going back to Plato and Aristotle, who took what is sometimes called an idealist or rationalist point of view. They believed that the only secure and trustworthy source of knowledge was the human mind itself, and so believed that all practical knowledge–astronomy, geometry, science as a whole–ought to be based on some combination of indisputable principles of reason and self-evident, therefore also indisputable, facts.

This was not what Democritus had thought, but his style of reasoning, at least to philosophers, had been out of style for many centuries. It rose up again with the help of the British empiricists, notably Bacon, Locke, and Hume. They recognized that certain kinds of knowledge neither were self-evident nor could be generated by reason alone. Some things, and this applies quintessentially to science, simply had to be found out through observation and experimentation. The gap in their philosophy, so it seemed at the time, was that they could find no wholly logical way of generating universal scientific laws from a finite number of seemingly arbitrary empirical facts. Science, it therefore seemed, was not a philosophically sound enterprise.

Scientists, in their pragmatic way, may have recognized this problem, but rather than jettison science because they couldn't figure out the right set of philosophical rules, they decided that since science seemed to be working just fine, philosophical rules were evidently not necessary. Boltzmann, like many physicists, relied on a sort of gut instinct that philosophers are apt to call "naive realism." Realism, because there's a profound belief that the world exists independently of humans and their thoughts, and naive because that untutored belief makes philosophers smile.

Maxwell, Gibbs, and their compatriots tended to work blithely in the pragmatic, empirical style and to disregard philosophy. Apart from scientific considerations, this was why the strictures of Mach and the energeticism of Ostwald found few takers in the English-speaking world. The Irish physicist George Fitzgerald derided energeticism as "a sort of well-arranged catalogue of facts without any hypotheses . . . worthy of a German who plods by habit and instinct."

Boltzmann, however, was in the midst of the Germans. Though of an essentially pragmatic nature in his science, there was enough of the rationalist in him that he couldn't stifle the urge to buttress his empirical methods with some sort of reasoned philosophical scaffolding. Starting in the mid-1890s, Boltzmann wrote a number of more or less philosophical essays in which he tried to portray his idea of what a theory of physics ought to be. In his 1895 letter to *Nature* defending the H-theorem, he said, "Every hypothesis must derive indubitable results from mechanically well-defined assumptions by mathematically correct methods. If the results agree with a large series of facts, we must be content, even if the true nature of the facts is not revealed in every respect."

This is as good and succinct a statement of a pragmatic philosophy for the working theoretical physicist as one can find. It is a step beyond truly naive realism. Boltzmann understood that theory corresponds only approximately to reality, and that as science progresses, the approximation gets closer. As to the question of reality, facts, and appearances, he observed that "we infer the existence of things only from the impressions they make on our senses. It is thus one of the most beautiful triumphs of science if we succeed in inferring the existence of a large group of things that mostly escape our sense perception."

This was the sticking point in Boltzmann's protracted dispute with Mach. The scientist must begin with what can be seen or otherwise detected, but in forming hypotheses goes beyond those direct perceptions to the putative existence of entities that cannot be seen or detected. Atoms are the perfect example. But to Mach it was the perceptions only that were the legitimate building blocks of science. Theorizing about what may or not exist beyond the horizon of the senses was not only speculation but actually antiscientific.

The gap between these two philosophies was unbridgeable. Boltzmann was hampered by his own honesty in recognizing that "there are no absolutes, especially no absolute truths." He realized that the true nature of reality, whatever that may mean, can never be finally established. In this respect, Boltzmann was again not an entirely naive realist; his naiveté was qualified by his understanding that even if there is a real and absolute world out there, scientists can apprehend it only by degrees, by a process of successive approximation. For Boltzmann this was not a fundamental deficiency of science, but a practical limitation on what the scientist can hope to achieve at any given time, and ultimately not a particularly serious problem. There's no need to establish an absolute definition of truth or reality in order to do science. One makes do. But to Mach this disability was a fatal flaw. If science cannot establish absolutes, what good is it? And rather than recognize that it's possible to make do, Mach tried to strip science down to a skeletal set of bare essentials, limited and meager but at least dependable.

Boltzmann knew, as any scientist must, that strict adherence to Mach's views–which he referred to as phenomenology, meaning that Mach would permit himself to rely on nothing but observable or tangible phenomena–would fatally hamstring scientific exploration: "Phenomenology believed that it could represent nature without in any way going beyond experience, but I think that this is an illusion. . . . The more boldly one goes beyond experience, the more general the overview one can win, the more surprising the facts one can discover, but the more easily too one can fall into error. Phenomenology therefore ought not to boast that it does not go beyond experience, but merely warn against doing so to excess."

Here was an impasse. Boltzmann argued that to make any progress, scientists were bound to speculate and hypothesize about a "reality" that lay beyond experience. Mach responded that if that's what scientists did, they weren't doing science any more. In the end, Mach was forced to deny that a lot of things scientists were doing in the late 19th century were really science. He was an extreme idealist: so important was it to him to maintain a kind of philosophical purity that he would deny that hydrogen and oxygen retained any independent existence once they had been joined together as H_2O, deny that under any circumstances it was possible to say whether cause preceded effect or vice versa. Mach and his followers valued this severe rationality more than they valued the ability to find things out.

Mach's philosophy did have one glaring weakness, however, which many besides Boltzmann pointed out. If scientists could depend only on direct experience and tangible phenomena, how could one scientist be sure that his or her experiences and observations were the same as everyone else's? Clearly, even for Mach, it was necessary that scientists should be able to agree on some things–that the blueness of the sky that I perceive is the same blueness that you see, too. But in Mach's stringent world, how to be sure of this? This is the peril of solipsism, of declaring so rigidly that one must believe only in the truth of what one knows personally to be true, that one cannot be sure of anything else.

Mach himself, to be fair, recognized the problem and strove to deal with it, but in the end resorted to the rather unsatisfactory declaration that by force of "irresistible analogy" it was possible to be confident that we all basically observe the world in the same way. But in this Mach was relying on the same sort of practical common sense that Boltzmann used as justification for believing in some sort of independent reality: we know it works that way because that's the way we know it works.

In an 1897 essay, Boltzmann retorted that if Mach were free to resort to an unspoken but "irresistible analogy" to presume that his sensations were essentially the same as anyone else's, then he, Boltzmann, was at liberty to make the same sort of assumption in believing that atoms are real, even though they lay beyond his personal experience. This is, in fact, a pretty good dig at Mach, who took no notice and merely observed that he had long been familiar with such "simple considerations."

DESPITE ITS EVIDENT limitations and restrictions, Mach's thinking had a powerful if ultimately transient influence on the development of physics. His philosophy had some worthwhile points. He emphasized the importance of sticking to experimental facts and observations as the final criteria for evaluating science, and he cautioned against extravagant theorizing. Even Boltzmann on one occasion praised Mach's dictum that "mathematics is economically arranged experience of counting," by which Mach seems to have meant that the core of mathematics is the synthesizing of simple arithmetic in as succinct a manner as possible. And Mach's warnings against hypothesizing to excess, of postulating the existence of things that have no possibility of being detected, is one that theoretical physicists today perhaps need to take more seriously and thoughtfully. The young Albert Einstein, learning his physics during the 1890s, was for a time under the influence of Mach's writing and later recalled that even "those who consider themselves opponents of Mach scarcely know how much they have taken from his way of looking at things, along with their mother's milk, so to speak."

Mach's influence in Vienna and his role in physics contained contradictory elements. The energeticists claimed Mach for their godfather on the grounds that they were formulating a system of physics and chemistry that took as its central element the plainly tangible, directly measurable phenomenon called energy. In his autobiography, Mach went so far as to claim that he had expounded the essential principles of energeticism in his 1872 book *The Conservation of Energy*. But Ostwald bristled at this and in his own autobiography settled the score: Mach "would not associate himself with energeticism. . . . he is much more critical than sympathetic." The reason for Mach's eventual antipathy was simple. Ostwald and his allies soon realized that energy principles alone were wholly inadequate to explain all of physics and chemistry, and to make their models useful they had to resort to additional theorizing and hypothesizing. Ostwald tried to devise different classes or categories of energy, including some that by physical standards represented quite fanciful distinctions. A number of the energeticists believed they were remaining faithful to Mach's ideals, but as they saw the need to elaborate their thinking, they began to develop theoretical notions that Mach found just as distasteful as atomism.

One physicist of rising influence found himself for a time outside both camps. Max Planck had argued consistently against kinetic theory. He believed that the laws of thermodynamics must be absolute, and that kinetic theory, with its probabilities instead of certainties, must therefore be wrong. But Planck opposed the energeticists with equal vigor. He, just as much as Boltzmann, saw that their aim of explaining all of physics in terms of energy alone was impossible, and that in many instances they simply did not understand thermodynamics correctly. Like Boltzmann, he bemoaned the influence of the energeticists, recalling later that "it was simply impossible to be heard against the authority of men like Ostwald, Helm, and Mach." And he regretted that Boltzmann appeared indifferent to his arguments against the energeticists: he regarded himself as "a second to Boltzmann, a second whose services were evidently not appreciated, not even noticed by him." But Planck wrote these words long after the battle was done, overlooking the fact that for the first half of his life, until the very last years of the 19th century, he had opposed Boltzmann's kinetic theory as vigorously as he opposed energeticism. As late as 1897, Planck was writing to a colleague that if there was a contest between thermodynamics in its classical form and the atom-based kinetic theory, then it was atomism that came off the loser. It "will be easier and more promising," he declared, "if one assumes the second law to be strictly valid (something which certainly cannot be shown from the kinetic theory of gases in its present form)."

Shortly afterward, Planck's opinion of kinetic theory underwent an abrupt reversal. He finally saw why the second law must inevitably be probabilistic, and thereafter became an active devotee of Boltzmann's views on thermodynamics. Throughout Planck's *Scientific Autobiography* runs a string of rather peevish references to Boltzmann's poor opinion of him, as if Planck cannot quite understand the reason. But his conversion to the cause of kinetic theory came rather late, and it is unclear whether Boltzmann overcame his antipathy enough to even notice that Planck was now on his side.

The circumstances of Planck's change of heart were dramatic. For many years, he had been applying his profound knowledge of classical thermodynamics to electromagnetic radiation–specifically, to the unsolved problem of explaining the spectrum of radiation in a closed cavity held at a certain temperature. A simple example would be to understand the heat and light generated within a furnace at an iron-smelting plant: Why does the glow change from red to orange and then even become bluish as the furnace temperature rises?

Since heat and light, after Maxwell, were now understood as electromagnetic oscillations of various wavelengths, the task was to link, through physical theory, the temperature of the furnace walls to the range of emitted wavelengths within. It proved an intractable problem. Planck and many others modeled the situation in an endless number of ways, but they always came up with predictions in which there was too much energy at short wavelengths (the blue end of the spectrum) and not enough at long wavelengths (in the red and infrared). Experimenters had succeeded in measuring the spectrum with considerable accuracy over a wide range of temperatures, and theorists could not explain what their colleagues saw.

Though he had until then always looked askance at Boltzmann's statistical methods, in the late 1890s Planck began to look again and to see something in them that he might use. In his memorable 1877 study, Boltzmann had devised a method for notionally dividing up the energy possessed by atoms into numerous allocations of some tiny unit of energy, and from that he had figured out how to count the probability of atomic distributions according to the way these energy units were divvied up among the atoms. This had led to his famous formula for entropy, $S = k \log W$.

Planck now hit upon the idea of using a similar trick to solve the radiation problem. Replacing the atoms in Boltzmann's theory with electromagnetic waves of different wavelengths, Planck imagined energy divided up into tiny units, and then allocated them among the different wavelengths. Using methods similar to those that Boltzmann had first devised, Planck mathematically sorted out all the different possible arrangements to decide which corresponded to a state of thermal equilibrium in the heat and light. Very quickly, he found his answer, and in 1900 he published one of the most famous papers in physics, explaining how the spectrum of radiation at a given temperature could be calculated with this trick of dividing energy up into little units.

Planck's solution to the longstanding radiation problem was quickly and widely recognized, but the significance of his method of solving it was not. This business of dividing the energy into units was perplexing, and Planck left unsaid what exactly he thought of it. Still, he finally had the answer that had for so long eluded him and the rest of the physics world.

In finding a solution to the radiation problem, Planck veered away from the principles that had guided his thinking until that time. He had always disliked kinetic theory because it denied the absolute truth of the second law of thermodynamics. Now, however, Planck found that by adapting Boltzmann's use of statistics and probability, he could at last get what he wanted. Planck is generally given credit as the man who, in 1900, brought the germ of quantum theory into the world, but it was only when he let himself understand Boltzmann's view of physics that he found the way forward. Boltzmann is plainly a grandfather to quantum theory.

But the genealogy of scientific credit is often a delicate matter. Planck sent Boltzmann a copy of his work, and in his Nobel Prize address of 1920 claimed that he had received, so to speak, a pat on the head from the older man: "it gave me particular satisfaction, in compensation for the many disappointments I had encountered, to learn from Ludwig Boltzmann of his interest and entire acquiescence in my new line of reasoning." But the circumstances of this blessing are left unstated, and no record from Boltzmann's hand indicates that he approved or even knew of Planck's work. On the contrary, Boltzmann had publicly pointed out errors and inconsistencies in some of Planck's earlier attempts to solve the problem. Moreover, Planck himself was reluctant to see the truly revolutionary nature of what he had done, and for a number of years he thought it would be possible to explain the division of radiation energy into little bits through purely classical physical reasoning. Many physicists besides Boltzmann remained unimpressed by Planck's new idea, except as an interesting if strange way of solving an irksome problem.

A few years later, however, other physicists showed that Planck's quantum hypothesis was wholly incompatible with classical physics, and it was arguably Einstein, a couple of years later still, who first took seriously the idea of quanta as independent elements, "atoms" of energy analogous to atoms of matter. Planck himself remained ambivalent for the rest of his long life.

Still, Planck's change of heart in the late 1890s caused him at last to understand and genuinely admire Boltzmann's work. Partly as a result of this reevaluation, his animosity toward Mach grew. At a 1908 scientific meeting in Leiden, Holland, Planck launched a harsh attack on Mach and his views. In a lecture titled "The Unity of the Physical World Picture," he made the point that although Mach had professed to offer a unified philosophy of physics, he had in practice balkanized the subject, dividing it into areas of unrelated phenomena and permitting no underlying or encompassing theories. He concluded by warning his audience against "false prophets" and declared, "by their fruits ye shall know them."

There was no doubt whom he meant. Some physicists came to Mach's defense; others agreed with Planck but thought perhaps he was being a little overdramatic. There is indeed in Planck's denunciation something of the tone of the spurned lover. In any event, Mach's influence was by that time on the wane.

IN A MORE IMMEDIATE sense, Mach's influence on Boltzmann had undergone an abrupt change 10 years earlier. In July 1898, while taking a train ride to visit his son Ludwig in the northern German city of Jena, Mach suffered a serious stroke, which left him paralyzed on his right side and unable for a while to speak. He survived, and with great willpower and determination continued to write. But his role as an active lecturer at the University of Vienna was done. He retired formally in 1901.

Even so, Mach's influence remained strong in Vienna. The future philosophers of the so-called Vienna Circle and a number of writers and journalists owed the formation of their views to him. In a city increasingly riven by nationalist and political conflicts, and held together, if at all, by the influence of the aging emperor and his court, Mach's philosophy of looking only at the surface of things and refusing to countenance deeper theoretical explanations may have provided a small sense of security.

One notable Machian was the precocious young Viennese poet Hugo von Hofmannsthal. In 1891, Viennese intellectuals were startled to discover that remarkable poems appearing above the nom de plume "Loris" came from the pen of a 17-year-old schoolboy who was forbidden by school regulations from using his own name. Hofmannsthal became a central figure in the artistic circle known as Young Vienna, and with some of his fellow believers he attended Mach's philosophical lectures. He was attracted to Mach's assertion that "the world consists only of our sensations." Surely this was a license for an observant and acute poet to understand the world as effectively as the scientists hoped to.

But Hofmannsthal suffered an artistic crisis at the ripe age of 25.

While he was young and unselfconscious, he was capable of taking in his unadulterated sense-impressions and turning them into elegant lyrics, but as he matured he found himself unable to resist the urge to turn his immediate impressions into something deeper, in effect to build a view of the world informed by his poetic insights. This kind of "theorizing," of course, was just what Mach forbade, so that his philosophy turned out to be as antithetical to poetry as it was to science.

Despite his infatuation with Mach's ideas, Hofmannsthal's native Viennese talent for muddling through proved more sustaining. He abandoned poetry, turned instead to drama, and became famous as the librettist for many of Richard Strauss's operas. Boltzmann had more difficulty letting go. It was never enough that he satisfied himself that Mach's philosophy was damaging or foolish; it continued to rankle that others still adhered to his opponent's point of view. Having begun the philosophical battle, he could not leave it unfinished, even if his better judgment suggested to him more than occasionally that the battle could never be won. Boltzmann remained at heart a scientific pragmatist for whom there could be no universal philosophy of science. Still, in the late 1890s, he wrote more on philosophy than on physics, taking little interest in the new physics that was emerging. He worked instead to defend atomism and kinetic theory against an array of critics, both real and imaginary. There was little stimulus for anything else in Vienna, where the university, though it continued to produce a few good students, had ceased to be a powerful center of physics research.

Word began to get around the scientific community that Boltzmann was once again restless and unhappy. In December 1898, Ostwald wrote to Boltzmann, after a lapse of some years since the debate over atomism versus energeticism in Lübeck. He wanted to raise confidentially the question of whether Boltzmann might be interested in a position in Leipzig, and he expressed hope that their scientific differences would not prevent them from working together.

Boltzmann wrote back quickly, in an almost effusive mood at

having heard from Ostwald again. "I saw from [your letter] that you are not personally angry with me," he exclaimed, and went on, "I will not hesitate to make clear what I have often said to German and Austrian colleagues, that I am not happy in Vienna, that I often regret having moved from Munich to Vienna . . ." (forgetting, of course, that when he was in Munich he had written to Loschmidt to say that he was no happier there than in "dear old Austria"). He complained of the lack of good students in Vienna interested in pure science, and said that if a position in Leipzig were available that was anything like the one he had enjoyed in Munich, he would jump at the chance to move.

But this plan quickly fizzled. There had been the possibility of a vacancy in Leipzig, but it was soon filled, and Ostwald had to write apologetically to explain to Boltzmann that there was nothing doing for the time being. In the spring of 1899, however, Henriette Boltzmann wrote to Ostwald to reinforce her husband's unhappiness in Vienna: "In Munich he was very happy and here he feels so extremely unfortunate." And she added that she felt a certain complicity, because it had been in part her urging that took him away from Munich and back to Vienna so that he could be guaranteed a pension. She was 10 years younger than her husband, and his health was far from robust; had they remained in Munich, she could see a dismal future for herself. But now Vienna was turning out badly.

Another opportunity arose. Ostwald let Boltzmann know in May 1899 that there was again the chance of an opening in Leipzig, although again other candidates were on the scene, particularly a local physicist by the name of Paul Drude. In June, Boltzmann heard from his old colleague Koenigsberger, whom he had met in Heidelberg as a young man on his first trip outside Austria. Koenigsberger asked Boltzmann if he knew of any good young physicists who might be interested in a position in Heidelberg. Boltzmann replied promptly, warmly recommending Drude. But this ploy bore no fruit, and in the meantime Boltzmann went to America for the summer.

Not until March of the following year did Ostwald write to Boltzmann with brighter news. Again, Boltzmann responded quickly, reiterating his dissatisfaction with Vienna. "I cannot conceal from you that I have become more and more dissatisfied with work in Vienna. Students dedicated to higher things are completely lacking. . . . I have as students only potential teachers for Gymnasium and worse, whose understanding of higher matters is almost nonexistent and who often come from Croatian or Slovenian schools.

"On top of that there is the shaky political situation in Austria. . . . If a possibility were really offered to me to escape from all this into genuinely fruitful activity suiting my inclinations, such as I had in Munich and so thoughtlessly gave up, then I would feel infinitely better and it would please me especially to owe this to you, since it would offer a striking example of the combination of scientific differences of opinion with the best personal friendship. Furthermore our differences of opinion might in many respects be bridged."

Boltzmann's concern over the political situation and his portrayal of the lackluster quality of his students were connected. Through the 1870s and 1880s Austria, and Vienna, had been ruled by generally liberal governments embodying the beliefs of affluent businessmen, the church, and the old aristocratic element. Suffrage was growing, but still limited, and the emperor's personal and political influence remained strong.

But the 1890s brought new forces into contention. In particular, a liberal-turned-demagogue by the name of Karl Lueger had been gradually amassing public support among disaffected shopkeepers and factory workers. He was the leader of what became the Christian Socialist movement, which was Catholic, nationalistic, and, after a while, anti-Semitic. In 1895, Lueger gained enough support in the city council to be elected mayor of Vienna, but the emperor refused to appoint him, fearing the intimations of mob rule that lay behind his politics. Two years later, Lueger's support had only grown, and even the emperor could not resist him. Lueger (who was coincidentally the same age as Boltzmann and had even been born in the same suburb of Vienna) became the focus of an odd collection of socialists, anti-Semites, and nationalists united mainly for the purpose of upsetting the status quo.

In that same year, 1897, new language ordinances in Bohemia and elsewhere led to riots in the streets of Vienna by German nationalists. As part of the continuing effort to placate non-Germans, Franz-Josef's chief minister, von Taaffe, had pushed through a rule making Czech an official language of the civil service in Prague. Since educated Czechs, by necessity, spoke German, while few Germans bothered to learn Czech, the effect of this rule was as much to disenfranchise German Bohemians as it was to embrace the Czechs. Nationalists in Vienna and elsewhere took to the streets, with the immediate result that Franz-Josef was obliged to fire von Taaffe. By this time, the emperor despaired of assembling a workable government of reasonable people, and ruled from then on by direct decree.

For a nonpolitical Austrian such as Boltzmann, the main effect of these upheavals was to make life in Vienna less pleasant and safe. He scorned the rising tide of nationalism, having suffered as rector of Graz through rioting caused by pro-German students, but his antipathy was based entirely on a desire to maintain the status quo, the quiet life.

Despite his complaints about poorly educated students from "Croatian and Slovenian schools," there is no record that Boltzmann showed any particular dislike of any of the numerous populations embraced by the Austro-Hungarian Empire. His revered teacher and mentor, Josef Stefan, was of Slovenian origin. But now Boltzmann seemed to find it simply a nuisance that non-German schools were not as good as the German ones. Others found this out more directly. The American physicist Michael Pupin, who visited Berlin and recorded his impressions of Helmholtz and others, was born a Serb, in a part of Franz-Josef's realm that had been given over to Hungarian rule. Pupin's father, formerly a loyal Austrian, angrily renounced his allegiance to the emperor and encouraged his son to seek prosperity elsewhere. Pupin tried Prague, but found himself looked down on there by both the Germans and the Czechs. So he went to America and became a success, and a symbol of Austria-Hungary's inability to treat all its citizens equally. Many others were not so fortunate or so enterprising; some of them were the ones who came to Vienna with inadequate educations, which Boltzmann found himself trying to remedy. He had no patience for the task. Vienna at the turn of the century was not the pleasant, prosperous, politically untroubled Vienna of Biedermeier days, even for a thoroughly middle-class speaker of German.

Both Boltzmann and Ostwald, as their letters indicate, made an overt effort to separate their scientific differences from their personal regard for each other. Ostwald appears to have decided that any unpleasantness arose from a part of Boltzmann's character for which he was to be pitied rather than blamed or attacked. Writing about Boltzmann after his death, Ostwald recalled him as being "a stranger in this world. Constantly occupied by the problems of science, he had no time or inclination to deal with the thousand trivialities that take a large part of the life of modern man, and which he handles instinctively. This man, whose mathematical acuity would not pass over the slightest inconsistency, showed in daily life the innocence and inexperience of a child."

In any case, bringing Boltzmann to Leipzig would be a coup for Ostwald professionally. He told Boltzmann that "the prospect of winning you has stirred the liveliest interest in the faculty. . . . a favorable outcome may be deemed very probable." Later that month, one Leipzig professor wrote to another that "rarely have members of the philosophical faculty wished for the appointment of a new colleague in such a unanimous way."

The local candidate, Drude, disappeared from the scene by taking up an offer at the University of Giessen in northern Germany, leaving the field open to Boltzmann. Ostwald was concerned that having pulled as many strings as he could to get a quick decision, he should not let Boltzmann wriggle away. An official offer was made in April, and Ostwald wrote to Boltzmann describing the excellent department he had built up in his 12 years at Leipzig, the eagerness of the students, the enthusiasm of the faculty. He concluded, "for me personally it would be a hard blow if you didn't come. . . . failure in this matter would undermine my whole reputation with the faculty and the ministry. Therefore, if you vacillate on their account, come on mine. You won't regret it."

Boltzmann set about obtaining his release from the Austrian ministry and wrote to Ostwald that his mind was made up. But this held true only for a moment. The possibility of Boltzmann's departure leaked into the Viennese newspapers, setting off agitation within both the Austrian ministry and Boltzmann's psyche. Now he wrote to Ostwald, "I have never concealed that I have a tendency toward nervousness, and so now it happens that I once again have appointment fever. . . . But hopefully I will pull myself together soon. . . . I will probably come to Leipzig soon."

This was in April 1900. Boltzmann visited Leipzig briefly and went immediately afterward to Dresden, the capital of Saxony, where an agreement for his new position was quickly made and signed. With everything apparently settled, faculty members in Leipzig were nevertheless eager to calm Boltzmann, in case he should still elude them. Stories of his vacillations were by now well known in the academic community: his accepting then turning down Berlin, his saying no to Vienna in order to stay in Munich, then leaving Munich for Vienna anyway the very next year.

One of Ostwald's colleagues wrote to Boltzmann at the end of April to congratulate him on his decision: "I hope you will be happy in Leipzig, so that you will not leave us again soon." As an added lure, the letter dangles the prospect of a new physics institute in Leipzig, for which "approval is very probable." In his reply, Boltzmann worries about his role in this new institute, complains about the quality of students that came to him in both Munich and Vienna, complains about "endless vexation" with technicians and assistants, about the difficulty he now has because of his eyesight in overseeing laboratory work. Replies from Leipzig hasten to assure Boltzmann that, in effect, whatever he wants to do is fine by them. Teach a little less; don't feel obliged to make any decisions about the new institute right now.

At this point, as had happened before, Boltzmann had committed himself to taking up a new position without having obtained permission to leave his present one. When he approached the Austrian authorities, they tried to change his mind, but at the same time Boltzmann's state of mind seemed evidently so fragile that even the Viennese minister of education, Hartel, wrote a long memo to the emperor trying to explain the eminent physicist's erratic behavior and pleading that his desire to leave Vienna, after all that had been done to bring him there and keep him, was not due to ingratitude or lack of patriotism but derived from idiosyncratic reasons of character. Hartel wrote plainly of Boltzmann's "almost pathological" desire for recognition and told the emperor frankly of the physicist's dissatisfaction with his standing in Vienna. What drives Boltzmann is "the feeling that here in Vienna he is not fully appreciated, and that elsewhere his professorial and scientific activities would be crowned with greater success; indeed," Hartel continued, "Boltzmann has repeatedly complained that he lacks competent students whereas, referring to his experience in Munich, he was able to gather around him a splendid collection of scientific apprentices from all nations, who could then throw still greater light on his glory as the Coryphaeus of science." (A Coryphaeus was the leader of a chorus in Greek drama.)

Whether moved or exasperated by this not entirely sympathetic portrayal of the famous physicist's plight, the emperor nevertheless released Boltzmann from his appointment in Vienna. Typically, Boltzmann himself now wavered. His health, physical and mental, had suffered ever since the first contacts with Leipzig. In May he had gone traveling away from Vienna for a few days, on doctor's orders, but had come back feeling not much different. In August, with the decision to go to Leipzig official and out in the open, Boltzmann found himself once more in the grip of neurasthenia (the 19th century's catchall term for all kinds of anxiety and depression), and went to a small sanatorium in the country. His ophthalmologist, Fuchs, visited him there and passed on reports to Minister Wilhelm van Hartel.

On his first visit, Fuchs arrived while Boltzmann was out walking and talked to the director of the sanatorium, who confided that Boltzmann had been in the grip of suicidal thoughts, so much so that he was never left alone. But the director assured Fuchs there was no sign of any "real psychosis" and that all would be well in due course. When Boltzmann returned and talked to Fuchs, however, he declared that he had changed his mind about Leipzig, because of the lack of good laboratory facilities, and that he intended to write to Hartel asking if his release from Vienna could be rescinded. Two days later, Fuchs went back to visit Boltzmann in the sanatorium only to find that he had left the establishment almost immediately after their previous conversation, and that he hadn't told anyone where he was going.

Hartel then received an undated letter from Boltzmann, referring to the wishes he had conveyed to Fuchs, but saying that the idea of remaining in Vienna had come to him only as a consequence of his mental disturbance, that he was committed to going to Leipzig, and that therefore he would indeed be leaving Vienna after all. The brief sanatorium stay had done little to calm Boltzmann's state of mind. An explanation is provided by another famous neurasthenic of the same era, Marcel Proust, who observed that "neurasthenics find it impossible to believe the friends who assure them that they will gradually recover their peace of mind if they will stay in bed and receive no letters, read no newspapers. They imagine that such a regime will only exasperate their twitching nerves."

Proust was equipped with a sophisticated sense of self-awareness, which a modern psychiatrist might characterize as an effective coping mechanism. Boltzmann had no such advantage. The physicist's true state of mind emerges in a plaintive letter he wrote to his wife while in the sanatorium:

> Dearest Mama!
>
> That we are separated so far is terrible for me. I have a guard here . . . and outside of him and some kindhearted nurses no one to talk to. The doctor could hardly care less about me.

> I sleep badly and am quite beside myself with misery. If someone would fetch me I would leave immediately. Alone they won't let me go. Please, Mama, come! Or arrange for someone else to come. Please have mercy and don't ask anyone else's opinions, but decide for yourself. Please forgive me everything!
>
> from your Lui

One way or another, Boltzmann appeared back in Vienna, but only briefly. In the fall of 1900, he and his family were in Leipzig, having slipped out of Vienna with no fanfare and barely a word to his colleagues and friends.

CHAPTER 10

Beethoven in Heaven

Shadows of the Mind

BOLTZMANN, LIKE MAXWELL, HAD A MINOR TALENT for versifying, and in his brighter moods would come up with bits of light doggerel to amuse his friends and colleagues. One recorded example of his poetical aspirations survives, from an unknown time in his life. Though Boltzmann called this piece "Beethoven in Heaven: A Choral Jest," it has a serious theme. He imagines himself ascending to heaven on his death and there encountering angelic music, which he, a lover of German romanticism, finds unsatisfactory:

> Bald bin ich dort: o reine weiche Klänge!
> Doch scheinen einförmig mir die Gesänge,
> was ich den Engelein auch nicht verhehle.
>
> Soon I was there: o pure, gentle sounds!
> But from the angels I could not conceal
> That in my ears their singing rang dull.

"You have a German soul!" an angel tells Boltzmann, envious of his musical gifts. The angel then explains that Beethoven himself, now resident in the celestial sphere, was persuaded by the Lord to compose something for the angelic choir. But that too Boltzmann finds unimpressive, and he tells Beethoven so. The composer's spirit agrees, and explains why his heavenly chorus is not up to terrestrial standards:

> Und weisst Du, was mir raubt des Schaffens Feuer?
> Der Töne mächstigster fehlt hier der Leier
> Und dieser mächt'ge Ton–es ist der Schmerz!
> Der so gewaltig klingt, der hallt wie Erz.
>
> Do you know what steals my creative fire?
> The most potent sound is lost from heaven's lyre
> And this mighty tone–it is man's pain!
> Which so fiercely resounds, which rings like iron.

Released from pain or suffering, in other words, even Beethoven is reduced to writing muzak. Creativity can't be detached from agony. This is a common enough if a rather adolescent sentiment, but as Boltzmann, in a state of agitation and unease, left Vienna for Leipzig, he could perhaps find relief only in the thought that his unhappiness was the fuel for something greater.

Maxwell is not known to have tackled any such weighty subjects in his verses, which were more often a matter of jokes and satires at the expense of his colleagues and himself. But he did come up with a few more reflective compositions, in one of which he pictures a weary student at the end of a hard day:

> Then, when unexpected Sleep,
> O'er my long-closed eyelids stealing,
> Opens up that lower deep
> Where Existence has no feeling.
> May sweet Calm, my languor healing,
> Lend me strength at dawn to reap
> All that Shadows, world-concealing,
> For the bold enquirer keep.

Maxwell imagines that the searcher's strength is renewed by the sleep, and that it is the respite from struggle, not the struggle itself, that takes the intellect forward. It was Maxwell, indeed, who stoically faced an early death. Boltzmann, wrestling with asthma, neurasthenia, numerous varieties of catarrh, failing eyesight, and a host of other aggravations great and small, was never to achieve repose.

As soon as he got to Leipzig, the enormity of his decision to move impressed itself on him. One professor there, accustomed to a more Germanic way of going about academic business, recounted Boltzmann's friendly and benevolent manner, in which he made no distinction between the professors and the students. But Ostwald, in a later sketch of Boltzmann's life and personality, reported that "despite the grateful reception by his new students he was beset in Leipzig by the most serious ailment that a professor can encounter: fear of lecturing. This man, who exceeded us all in the acuity and clarity of his science, suffered dreadfully under the unbearable worry that his mind and memory would desert him suddenly during a lecture."

He was uneasy and unhappy. "Anyone who got to know the ill and depressed Boltzmann in his Leipzig period would hardly believe" that this was the same brilliant man of earlier, healthier times, another colleague recollected. Scientifically, Boltzmann achieved nothing of note. He published little, just a few short remarks mixing physics and philosophy. And he paid little attention to what anyone else was doing. Early in his stay in Leipzig, he must have received from Max Planck a copy of the famous 1900 paper describing the new "quantum" method for explaining the radiation spectrum. Despite Planck's subsequent claim, Boltzmann made no public mention that he noticed this work or found it important.

A little later, Boltzmann was also sent an early paper by the young Albert Einstein, but his response to that, if any, likewise remains unknown. He was invited to go to Yale for the 200th anniversary of the university's founding, but wrote back to Gibbs in a shaky hand declining the invitation for reasons of poor health. Another pleasure in his life that began to slip away was his piano playing. Ostwald's daughter described later how musical evenings at her family's house on Saturdays seemed to lift Boltzmann's mood. His eyesight was becoming so poor, however, that he would have to balance two or sometimes three pairs of glasses on his nose in order to read the music. Sometime in 1901, Boltzmann wrote to Ostwald begging off the usual invitation. "I really haven't been feeling well; I suffer such nervous agitation that I must on that account leave off our musical Saturday evenings for the time being."

In the summer of 1901, he was taken by his son, Arthur, on a Mediterranean cruise, for relaxation and recuperation. After sailing from west to east, visiting Lisbon, Algiers, Malta, Athens, and Constantinople, they had to turn back because of an outbreak of the plague in Odessa, on the Black Sea. The ship lay in quarantine. Father and son played chess. The weather was hot and uncomfortable, especially for the corpulent and now almost 60-year-old physicist. "Papa sweats and swears all the time," Arthur wrote home to his family.

No sooner had Boltzmann left Vienna than the thought of returning was in his head. Hartel, the Austrian minister, kept in touch. As early as February 1901, a bare six months after he had arrived in Leipzig, Boltzmann was writing back to Hartel expressing a hope that in the not-too-distant future a return to his homeland might be possible. As much as Boltzmann wanted to return, Vienna wanted him back. The Austrian universities already felt themselves to be slipping down in reputation and achievement compared to their German counterparts, and Boltzmann's departure was a further insult. Whether Boltzmann had any more scientific accomplishments within him was beside the point; it was not so much the man himself as his reputation, his past achievements, that Hartel wanted to bring home.

By June of 1901, Boltzmann was already in possession of an official letter offering him a position in Vienna; he replied expressing his regret at leaving Austria and his desire to return. Still, and needless by now to say, certain negotiations had to be undertaken. Alhough he had openly admitted his unhappiness in Leipzig, Boltzmann could not desist from haggling over salary and other details. The physics department in Vienna, in recognition of its dilapidated state, both physically and intellectually, would be rehabilitated, and Boltzmann was anxious to be assured that if he were not awarded use of the department's old residence on Türkenstrasse there would be some additional compensation coming his way.

He did promise from the outset, however, that once he returned to Vienna he would never again seek employment elsewhere. Hartel worked to assure the emperor that Boltzmann's abrupt departure from Vienna was the result of mental stress and derangement and, somewhat contradictorily, that once the physicist was back in Vienna again, his mental state would be repaired and he would stay put. He used the same phrases to describe Boltzmann's motive for returning to Vienna that he had recently used to explain his departure. It was Boltzmann's "almost pathological ambition" that had to be satisfied, his need to be recognized as "a Coryphaeus of the first rank."

Negotiations dragged on into 1902. Boltzmann's unhappiness festered. George Jaffé, a student in Leipzig, recalled that Boltzmann "must have felt that he was on the defensive and was somehow overshadowed by the brilliance of his friend and opponent [i.e., Ostwald]; at least so it seemed to us who were in the energetics camp. As a matter of fact Boltzmann was very unhappy in Leipzig. He was homesick and longing for his native Austrian mountains. His depressive mood must have reached a climax for he made an attempt at his own life while he was in Leipzig."

Mention of this reported suicide attempt comes only at second hand, from Jaffé and in stories handed down through Boltzmann's family, but it is known that in Leipzig he put himself for a time under the care of a local psychiatrist. Boltzmann's own letters testify to his misery in Leipzig and to his again fragile state of mind. Perhaps, he speculated, it was the swampy climate, so different from the clear air and hills of Vienna; perhaps the food; perhaps the "north German protestant way of life." By early 1902, he confided in another letter, he was "somewhat nervous and confused."

In June of 1902, Boltzmann pleaded ill health in order to be released from his servitude in Leipzig. Meanwhile, Hartel was working on the emperor to get the errant physicist back into Franz-Josef's good graces. Boltzmann's scientific eminence–the very thing that caused him to be dragged this way and that by offers from universities across Europe–was not in any doubt. Hartel, instead, was blunt about Boltzmann's failings and difficulties. From his early days in Graz, he had been eager to acquire a reputation and a position to go with it. He recalled Boltzmann's indecisiveness, the ease with which flattering offers would pull him in, only to trap him. He described his neurasthenic sufferings, the advice he had sought out from doctors everywhere he went. Finally, he assured the emperor that this time Boltzmann was home for the duration.

In one possibly apocryphal story, Hartel asked the emperor to imagine a wonderful but flighty dancer who ran away from the Vienna theater but then wanted to return: would the emperor take back such a favorite? Franz-Josef, who for many years had maintained a close relationship with a famous Viennese actress, Katharina Schratt, reportedly laughed, and consented to Boltzmann's reappointment.

Emperor Franz-Josef had no pretensions to being an intellectual. Like most Viennese, he enjoyed light theater and musical entertainment, but when he was obliged to attend more serious artistic events and openings, he would make a cursory visit, tell his hosts, "It was lovely; it pleased me very much," and depart. He certainly knew that Boltzmann was a famous and eminent theoretical physicist, but he could not have said what Boltzmann was famous for. Having signed the papers that brought Boltzmann back to Vienna, the emperor's purpose was achieved. Keeping Boltzmann happy, now that he was home again and had agreed to stay home, was of less concern.

The Institute of Physics to which Boltzmann returned in the second half of 1902 was becoming a run-down, neglected place. The Türkenstrasse building itself, the "temporary" home of the institute for some 30 years, was cramped and falling apart. Lise Meitner, a student at the time who was later to become famous for her role in discovering nuclear fission, remembered thinking that "if a fire breaks out here, very few of us will get out alive." The quality of physics teaching was still high, but apart from Boltzmann himself, there was no truly outstanding physicist. There was talk of spending money and putting up a new building, but as always, progress was excruciatingly slow. Soon after his return, Boltzmann and Franz Exner, a younger physicist to whom most of the administrative responsibilities had fallen, wrote to the authorities urging haste, but it was not until 1912 that a new physics institute finally came into being. (This Franz Exner was the son of the Franz Exner who had inspired Josef Loschmidt to a career in physics many years before.)

Boltzmann returned to Vienna as merely an ordinary professor of physics, not a department or institute leader, and his old rooms at the institute were now occupied by Exner. In any case, Boltzmann seemed to have little interest in any new physics. He continued to lecture and give seminars from time to time, but had few regular teaching duties. Meitner recalled his teaching: "His lectures were the most beautiful and most stimulating that I ever heard. . . . He was himself so enthusiastic about all he was teaching that we left every lecture with the feeling that an entirely new and wonderful world was being opened to us." But another young physicist, Paul Ehrenfest, noted in his diary his annoyance when Boltzmann's seminar was canceled, and recorded that when he had mentioned some new work on Becquerel rays–that is, radioactivity–Boltzmann apparently had no familiarity with the matter and no opinion on it.

Physics had moved on. The debate over kinetic theory had stagnated. Young physicists, for the most part, took the existence of atoms for granted without having any inclination to delve into the abstruse discussions that had engaged the followers of Boltzmann, of Ostwald, and of Mach for the previous decade or more.

One of the cornerstones of Boltzmann's career, moreover, was coming to be associated with another name. In April 1903, Willard Gibbs had died suddenly and unexpectedly, but just the year before he had completed a book, *Elementary Principles in Statistical Mechanics,* which established both the name and the style of a branch of physics that Boltzmann had in fact done much to create. In his book, Gibbs analyzed in his characteristically complete and utterly logical style the treatment of large physical systems as "ensembles" of microscopic elements. The energy, temperature, entropy, and other familiar thermodynamic properties of any ensemble, Gibbs showed, followed directly from the arrangement and statistics of its microscopic constituents. Changes of thermodynamic properties that resulted when an ensemble was squeezed, heated, frozen, or otherwise pushed around, or that occurred when chemical reactions intermixed the ingredients, could all be calculated in an elementary way from the behavior of the microscopic components.

As he had always done, ensconced in New Haven away from the philosophical debates of the German physics world, Gibbs strove to make his arguments rigorously logical, thoroughly comprehensive, and free of any but the most general assumptions about the physical identity of the "components" that made up ensembles. His book, like all his work, was a little arid in its intellectual style, but it came to be recognized as the embodiment of its subject matter: this was exactly how statistical mechanics ought to be presented.

Nevertheless, what Gibbs had achieved was in large measure a perfection of ideas that Boltzmann had sketched out a quarter of a century before. It was he who had introduced the idea of thinking of a physical system as an arrangement of microscopic elements, whose overall properties emerged from an application of statistics and probability theory. This was the great innovation of Boltzmann's 1877 paper, in which he had established the entropy formula S = k log W, W being the number of possible internal arrangements within what Gibbs would now call an ensemble.

Moreover, Boltzmann had published a paper in 1884 that more directly anticipated Gibbs. There he had explicitly discussed collections of states of gases, each state corresponding to a particular distribution of energy among its atoms, and he had analyzed the stability of these sets of states under various conditions. This was very similar to Gibbs' later depiction of ensembles, and a close correspondence exists between the families of states that Boltzmann described in 1884 and the different types of ensembles that Gibbs defined almost two decades later.

Still, the assignment of scientific priorities is rarely an easy matter. Maxwell, with his demon, had been the first to see the role of probability in the second law of thermodynamics, but it was Boltzmann (prodded by Loschmidt and others) who made these ideas quantitative. In the 1870s, Boltzmann and Gibbs, in their different ways, analyzed large systems in terms of the statistics of their component parts, bringing out the idea of thermal equilibrium as the most probable arrangement of such a system. Boltzmann's 1884 paper takes this idea further, laying the foundation for what is now called, after Gibbs, statistical mechanics. Maxwell also had an inkling of what was to come. His last paper on kinetic theory, written in 1878, contains the first hint of what would later become known as a statistical ensemble: rather than imagining a single system moving endlessly from one state to another over the course of time, he thinks instead in terms of a large set of equivalent systems placed side-by-side, each one exhibiting one of the possible states available to all. This change in perspective, though seemingly a matter of abstruse technicality, was crucial to the formulation of statistical mechanics in the widest sense.

Maxwell died the year after this last publication, and it was Boltzmann and then Gibbs who developed the subject. The standing of Gibbs' book *Statistical Mechanics* as a monument in the annals of science is well deserved, but it is also true that Boltzmann's earlier and less complete version of the subject is undeservedly neglected.

What Boltzmann himself thought about all this is unknown. Gibbs' book appeared in 1902, the year of Boltzmann's final return to Vienna, and was translated into German in 1905. Surely he must have known of it, but in his decreasing role as a lecturer in physics, he seems mainly to have described the system he set out in his own *Lectures on Gas Theory* and paid little or no attention to other methods of analysis. His few written statements about Gibbs seem mainly designed to portray him as a covert atomist, contrary to the attempts of Ostwald and the rest to depict him as an energeticist. In the end, it would hardly be a surprise if Boltzmann, who through his entire scientific career strove to create a subject as he thought it should be created, never made much of an effort to understand thermodynamics from Gibbs' point of view.

SCIENCE ASIDE, Boltzmann was undoubtedly happier in Vienna than he had been in Leipzig. He and his wife bought a cottage away from the city center. "The little house we bought in Vienna suits me and my family very well, and I hope that by spring and summer when we can sit in the garden it will please us even more," he wrote to Arrhenius in a letter dated New Year's Eve, 1902.

He acquired a new assistant, the young physicist Stefan Meyer, with whom he got on well. Meyer wrote down his recollections of Boltzmann in 1944, the centenary of his birth, and some anecdotes show that Boltzmann's good humor could still rise to the surface. On one occasion, Meyer was putting some of the less frequently used scientific journals away in an upper floor of the library, and Boltzmann came in and asked him what he was doing. Meyer explained. Boltzmann asked him what was the particular journal he had in his hand. "Dingler's journal," replied Meyer. "Now, indeed," said Boltzmann, "Dingler's journal is seldom used." That's why he was putting it away, Meyer explained. "Now," replied Boltzmann, "Dingler's journal really will be seldom used!"

Boltzmann's social awkwardness had not moderated over the years. He would greet visitors by booming out, "What do you want?"–by which, Meyer explains, he meant what anyone else would mean by, "How may I help you?" or "Please take a seat." On one occasion, Boltzmann and his wife invited a number of people for lunch at their cottage. When the guests arrived he and his daughter Henriette were standing there saying not a word to anyone. To break the oppressive silence, another guest started making the introductions, at which point Boltzmann abruptly said, "Now, I must go and see where my wife is," and disappeared.

The Boltzmanns' middle daughter, Ida, had stayed behind in Leipzig to finish school, and Henriette's letters to her provide a startling picture of her husband's ups and downs. She was at first pleased with the move to the cottage and that they were no longer living in the university accommodations at the physics institute, since Boltzmann had to get out and walk a little. Early in 1903, he was worse again, with a kidney problem. But then he seemed to get better. After that, though, "Papa is neurasthenic the nights before lectures," Henriette reported to Ida. "The doctor says he should lose weight." Henriette went behind her husband's back to get medicines, but then he wouldn't take them anyway. "My confidence in the future has been severely shaken. I thought things would turn out better," she lamented.

Boltzmann's intellectual energies were briefly revived by a new task. Ernst Mach, though he had struggled mightily to overcome the disabilities caused by his stroke, had formally retired from the university in 1901 and had gone to live near Munich. No one had been teaching his philosophy class in the meantime, and although many of the other faculty members wanted a "real" philosopher to take over, the desire to have someone teach a class in scientific thinking in particular caused Boltzmann to volunteer himself for the job. In May 1903, it was agreed that he should teach the class that Mach had taught, although he was never formally appointed as a professor of philosophy. This turn of events was greeted with a mixture of condescension, amusement, and dismay, and he himself, in his inaugural lecture as a philosopher of science, set off great merriment by musing out loud, in his high-pitched voice, "How do I come to be teaching philosophy?"

It was in this lecture that Boltzmann recalled how Mach had interrupted a discussion some years before to declare, "I don't believe that atoms exist!" From that incident sprang his interest in philosophical matters, he explained, and he made conscientious attempts to understand Mach's point of view and the various analyses of other philosophers over the centuries. But the undertaking baffled him. He tried Hegel, he said, but "what an unclear, senseless torrent of words I was to find there!" Schopenhauer was no better, and even in Kant "there were many things that I could grasp so little that judging by his sharpness of mind in other respects I almost suspected that he was pulling the reader's leg or even deceiving him."

Boltzmann's first philosophical lecture was a mix of broad jokes and belligerence, and he kept his audience amused with his act of a simple man trying to understand the ruminations of all these great minds. Nevertheless, the underlying tone was one of baffled frustration: what on Earth were these philosophers trying to say, and why?

When he first tackled philosophy as a teacher, Boltzmann perhaps thought he would make some general observations, on the strength of his own experience, and proceed to some sort of rough-and-ready working philosophy that would be as much as any scientist ought to need. But he soon bogged down. He was trying to make reasoned statements in a subject that, at heart, he thought basically futile. He had recently begun a correspondence with Franz Brentano, a former philosophy professor in Vienna who had retired to Florence; it was his departure from Vienna in 1895 that had opened a position and allowed Mach to return from Prague. Boltzmann tried out some of his ideas on Brentano, who had the grace and sometimes the courage to take his philosophical striving seriously. But Boltzmann's fundamental skepticism was never far from the surface. "Is there any sense at all in breaking one's head over such things?" he asked Brentano in 1905. "Shouldn't the irresistible urge to philosophize be compared to the vomiting caused by migraines, in that something is trying to struggle out even though there's nothing inside?"

Boltzmann's first two or three philosophy lectures attracted audiences in the hundreds and notices in the Vienna newspapers. News of his success even reached the emperor's ears, and at an imperial dinner to which Boltzmann had been invited, Franz-Josef told him personally how pleased he was that he had returned to Vienna. At the end of 1903, Henriette reported to her daughter that Boltzmann was "lively and animated . . . completely changed as if by magic."

But neither the success of his lectures nor Boltzmann's enthusiasm lasted very long. Within a month or two he was beset again by depression, anxiety, and other health problems, and weighed down by the burden of composing his philosophical thoughts.

In February 1904, Boltzmann was 60 years old, and his colleagues assembled for him a "*Festschrift*" of scientific papers by 125 scientists from all over the world. One participant hoped his contribution would "provide a small moment of happiness for our Boltzmann." Another recalled later that on the evening of his birthday, Boltzmann fell to reminiscing and explained that he had been born in the night between Shrove Tuesday and Ash Wednesday–a timing that he thought accounted for his lifelong swings between joy and misery. Behind the scenes, Boltzmann's name was being proposed to the Nobel committee for a physics prize, most notably by his former critic, Max Planck. But any joy deriving from his birthday celebrations was short-lived. By April he was begging off his lectures again, pleading a "powerful nervous depression."

Although Boltzmann was on his word of honor to the emperor not to leave Vienna for an academic position elsewhere, he was not under any injunction to leave the city for other reasons. In the summer of 1904, he sailed for the second time to the United States, mainly to attend a scientific meeting that was taking place in conjunction with the World's Fair that year in St. Louis. He went with his son, Arthur, on the ship *Belgravia,* which he says in a letter home was "very inferior and has many inconveniences." After having agreed to undertake the journey, he complains about it incessantly. From Washington he writes, "We have seen nothing new. . . . Recent days have not been happy ones at all." On the way back, just embarking from New York, he says, "I am ill and if this terrible melancholy which grips me at the moment does not subside in Vienna I shall be unable to deliver my lectures."

The highlight of Boltzmann's visit in 1904 was his participation in yet another scientific debate over atomism, with Ostwald once again taking the opposing view. One witness to this discussion was the young American physicist Robert A. Millikan, who was later to win the Nobel Prize for his experimental determination of the charge of the electron. In his autobiography he wrote, "The amazing thing now is that this question could be debated at all at that time, and that outstanding men like Ostwald and Helms [sic], and even the brilliant philosopher Ernst Mack [sic], could at that epoch" oppose atomic theories. To an audience of young New World scientists, this debate must have seemed an intrusion into their fresh universe of stale air from the Old World's attic.

BACK IN VIENNA in the fall of 1904, Boltzmann resumed his philosophical lecturing, but in desultory fashion and with little of the pleasure or enthusiasm he had at first imagined the subject would bring him. In January of the following year, he made another attack in the form of a lecture to the Vienna Philosophical Society that was originally to be delivered under the striking title "Proof that Schopenhauer is a stupid, ignorant philosophaster, scribbling nonsense and dispensing hollow verbiage that fundamentally and forever rots people's brains." He protested, to no avail, that he had adapted this title directly from Schopenhauer himself, who had used precisely these words to attack Hegel. This, he meant to say, was how philosophers go about their intellectual disputes. Nevertheless, he was persuaded to give his talk the more acceptable rubric "On a Thesis of Schopenhauer's." In his lecture, Boltzmann mentioned anyway what he had first hoped to use as the title, and added that Schopenhauer had a way of expressing himself that might have been "associated in the past with fishwives but which today would be called parliamentary."

For all the crudeness of his attacks, Boltzmann wanted to be taken seriously. He sent a copy of his Schopenhauer lecture to Brentano in Florence, asking for a frank opinion. Brentano replied guardedly, trying to persuade Boltzmann that his attacks were not quite so brilliant as he supposed. The lecture "steps very rashly from one point to another. Many of your criticisms show lack of careful judgment."

Boltzmann invited himself to stay with Brentano in Florence for a while so as to have the benefit of direct philosophical discourse with the man he now saw as his guide and adviser. Whatever Brentano's feelings about Boltzmann's philosophy, his sympathy for the man was heroic, and he was host to Boltzmann in Florence for about the first three weeks of April 1905. Boltzmann's express intent was to talk things over with Brentano and then compose a substantial philosophical monograph expressing his views. The book never materialized. Judging by a brief letter of thanks written after he was back in Vienna, Boltzmann had listened but not changed his opinions in any substantial way. Then he proceeded to tell an amusing tale about how he had difficulty finding a hotel in Venice on his trip back from Florence, and after following a guide down a narrow alley had almost got wedged tight, and had to wiggle out sideways. Boltzmann resisted Brentano's attempts at philosophical guidance as much as he ignored his doctor's instructions to lose weight.

No sooner was he back in Vienna than he was making travel plans again, this time for his third visit to the United States. The University of California at Berkeley had instituted a summer session to which it invited prominent European scientists as guest lecturers. Boltzmann's name had come up, and despite the endless grumbling about his previous visit to America in 1904, he had accepted. On June 8, 1905, Boltzmann took the train to Leipzig, stayed there with friends for a few days, sailed from Bremen on the north German coast a couple of days after that, reached New York on June 21, and set off by train across the continent.

Unlike his previous visits, this time he traveled alone, later writing an account of his journey and of California in a piece he called

"The Journey of a German Professor to Eldorado." Characteristically, his tale veers between enthusiasm and griping, between rhapsodizing over the sights and people of the New World and grumbling over their shortcomings. The journey by sea sends him into spasms of romantic delight. "See that ship over there! Now it's swallowed by the waves! No! Instantly her keel rises up in victory. . . . On a few exceptional days the sea adorns itself in a dress of the finest ultramarine, a color at once dark and luminous, and trimmed with milk-white foam as if with lace. I wept at the sight of that color; how can a mere color make us cry? And then the moon's glow or the sea's brilliance in the pitch-black night." By contrast, the scenery on land, as he is speeding from New York to California, mostly fails to impress him, although he allows that the Sierra Nevada are not without merit; they are not as picturesque as the Austrian Alps, but more magnificent in their height and extent.

As they do for many a querulous traveler, food and drink feature prominently in Boltzmann's account. Before leaving Vienna, he has a last sturdy meal of roast pork and potatoes at the restaurant in the train station and loses track of how many glasses of beer he swallows. On the ocean liner, a German ship, he can enjoy familiar cuisine, but in America, and especially in California, he is in a gastronomic new world. Most testing for him is that Berkeley turns out to be a dry town, and his stomach rebels after drinking the local water, which, he complains, is collected in rain buckets. When he inquires of a colleague how one may obtain beer or wine, the reaction he gets is as if he had been asking about quite another sort of establishment. But he is directed to a wine shop in Oakland and learns how to smuggle bottles of wine back to Berkeley. Thereafter, as Boltzmann put it, "the road to Oakland became very familiar to me." Even so, the necessity of stealing wine into the room where he was staying and drinking only when he was alone, after dinner, began to make him feel like "depravity's slave."

Invited to dinner at the hacienda of Mrs. Hearst (mother of William Randolph and a great benefactor of the university at Berkeley), he is first offered fresh blackberries, which he declines, and then melon, which he must also refuse. After that, he turns away from "an indescribable paste made from oatmeal, which in Vienna might be used to fatten geese, except that I doubt true Viennese birds would touch the stuff." But eventually he is brought chicken and other cooked items, and can dig in.

But America is not all deprivation and hardship. The Berkeley campus is beautiful, and a fireworks display over San Francisco on the Fourth of July, as well as the boisterousness of the people, brings home the exuberant belief of Americans in their country and their ideals. He visits the great telescopes at Lick Observatory, near San Jose, completed in 1888 with funds from the financier James Lick, whom he calls an idealist and a millionaire. "Happy the land where millionaires hold ideals and idealists become millionaires!" All in all, he concludes, "America will accomplish great things in the future. I believe in these people."

Of the reason for his stay in Berkeley, which was to deliver a course on theoretical physics, Boltzmann had little to say. He had been told beforehand that he could lecture either in German or in English, German then being fairly widely understood by scholars of all sorts, but he had decided his English was up to the task, and had even taken some conversational classes in Vienna before leaving. His eagerness was dampened when he first tried to ask about lunch on the train. "Lernch, lanch, lonch, launch, . . ." he tries, before finally getting through. "And now I have to give thirty lectures in this language?"

But once he started lecturing he was pleased with his fluency. He manages tongue-twisters like "blackboard" and "chalk" with ease, and smoothly utters words such as "algebra," "differential calculus," and the like.

His listeners came away with a different impression. His English was idiosyncratic. "Dass ist the truth!" he would exclaim, when the right word deserted him. A flavor of his headlong style can be had from a letter he wrote in English to Lord Kelvin. After discussing some technical matters he finishes by saying, "I were extremely glad, if You had the complaisance to write to me, if I succeeded by my small knowledge of the English language to exprime clearly enough my ideas." An observer in Berkeley reported that his English was "somewhat deficient, to put it mildly." Faculty members and students at Berkeley found Boltzmann's visit something of a disappointment, though Boltzmann, in his oblivious way, seems to have had a fine time.

The journey home was arduous. Through the dry state of North Dakota Boltzmann had to bribe a railway attendant, at considerable expense, to bring him wine. When a strike by telegraphers delayed the trip, Boltzmann was thrown into a rage, and was further agitated by the phlegmatic calm with which Americans accepted such setbacks. Finally, he was overjoyed to reach New York and board a German ship. The food calmed his stomach, and he "drank not a drop of water, not much beer, but therefore all the more noble Rüdesheimer [wine]," he wrote in his account. An ocean voyage is the ideal venue for such enjoyment, he added, since "if one totters a little, it can all be blamed on the rolling of the ship."

CHAPTER 11

Annus Mirabilis, Annus Mortis

Einstein Rises, and a Man Falls

"NOW THERE IS ONLY THE TRIFLING train journey from Bremen to Vienna, a smart ride in a Viennese fiacre, and I am at my house. . . . the most wonderful moment of the whole expedition is the moment when one reaches home again."

So Boltzmann concluded his account of his California adventure, and indeed there were reasons his return to Vienna might have been a triumph. The year 1905 has rightly been called an *annus mirabilis* for physics; over the course of that year a young physicist named Albert Einstein, just 26 years old, published four papers that between them set the stage for much of physics in the 20th century. The two later works embodied what is now called the special theory of relativity, the subject most often associated with Einstein's name. The two earlier ones were arguably even more revolutionary. In one, Einstein clarified the meaning of Max Planck's five-year-old explanation of the radiation spectrum, establishing, as Planck had not done, the physical reality of what are now called photons: quantum particles of light. In the other, he presented a simple explanation for a phenomenon that had puzzled scientists for close to a century, and in doing so provided almost a direct proof of the existence of atoms. The first paper explicitly used methods that Boltzmann had devised decades earlier; the second showed that Boltzmann's lifelong confidence in atoms was well founded. Einstein thus borrowed from Boltzmann and immediately repaid the debt.

Einstein's name was not well known in 1905, but neither was he a novice. He had begun to publish in 1902. Born in 1879, Einstein had grasped without hesitation the importance for physics of Boltzmann's embrace of statistics and probability. Unlike the physicists of earlier generations, he had no ingrained preconceptions to set aside. While still an undergraduate, Einstein wrote to Mileva Maric, a fellow physics student whom he was to marry a few years later, "Boltzmann is quite magnificent. . . . I am completely persuaded by the correctness of the principles of the theory, that the question is really about the movement of [atoms] according to certain conditions." He was reading the *Lectures on Gas Theory,* Boltzmann's own account of his life's work in science. The dense and detailed two-volume monograph was only a few years old, and its intricacies defeated many of Boltzmann's contemporaries. It was not typical reading for the average undergraduate.

His interest in Boltzmann's statistical perspective led Einstein to the first original work of his career. In the three years before 1905, he published three elegant analyses that together made many of the arguments that Gibbs had assembled at the same time in his *Statistical Mechanics.* His work combined the best aspects of both Boltzmann and Gibbs. Like Gibbs, Einstein was possessed of a ruthless logical efficiency, baring the subject to its elementary foundations, stripping away what was unnecessary until he had a simple structure, powerful precisely because of its simplicity. But like Boltzmann, and rather unlike Gibbs, he also had an almost visceral feel for the physics itself, able to divine, one might say, that things had to work a certain way before he had captured in mathematics the reasoning that would get him there.

These early endeavors, steeped as they were in Boltzmann's thinking, helped prepare Einstein for the revolutionary year of

1905. The first of his papers, delivered in March, demonstrated clearly the true nature of Planck's explanation of the radiation spectrum. In 1900, Max Planck had borrowed heavily from Boltzmann when he divided up the energy of a volume of electromagnetic oscillations into smaller units in order to provide a theoretical derivation of the spectrum of radiation emitted by an object maintained at some temperature. But the meaning of this division into units of energy was enigmatic. Boltzmann had segmented the energy of a gas into units in order to obtain his entropy formula, but this divvying up was quite clearly a mathematical device. There was no implication that the energy actually existed only in discrete lumps. Indeed, it was an elegant part of Boltzmann's argument that when he had finished the calculation, the actual size of the units was unimportant, as long as it was sufficiently small to meet certain mathematical conditions.

This was decidedly not the case in Planck's analysis of radiation, and therein lay the puzzle. The magnitude of the energy units entered explicitly into the radiation formula Planck derived, and to get the formula to agree with experiment, they had to be of a certain size. Planck believed at first, and for some time afterward, that this was in some way merely an oddity of the manner in which he had done the calculation and not that it had any fundamental physical meaning. Undoubtedly, in 1900, Planck did not believe he had proved that radiation energy must come in small units–atoms of energy, so to speak.

Over the next few years, Planck and others persisted in thinking that some explanation, in terms of well-established physical principles, would in due course be found for these apparent "quanta." The word *quantum,* incidentally, acquired its modern scientific meaning only gradually. In German it is an ordinary word for "quantity," and Boltzmann had even used it in the title of a paper he published in 1883.

In the early years of the 20th century, several physicists came to realize that any explanation of the radiation spectrum based on Maxwell's theory of electromagnetic waves allied with basic principles of thermodynamics must predict a radiation spectrum inconsistent with experiment and therefore also inconsistent with Planck's formula. There was, in other words, something quite new implicit in what Planck had achieved.

Einstein's thorough reading of Boltzmann gave him the tools to see what that new thing must be. Like Planck, he borrowed Boltzmann's techniques, but he went further. The essence of his first great achievement of 1905 was to show that treating each "quantum" of radiation energy explicitly as a physically independent entity yielded formulas for the energy and entropy of a volume of radiation that agreed exactly with a derivation of these quantities along traditional thermodynamic lines (these formulas also having Boltzmann's name on them, in the form of the Stefan-Boltzmann law). The implication was clear and striking: just as a physical gas is composed of individual atoms, so a "gas" of electromagnetic radiation appears to be composed of distinct quanta. The division of energy into small units was not merely a mathematical trick but in truth represented a new and astonishing discovery about the physical nature of electromagnetic radiation. The quanta truly are atoms of energy.

Einstein then showed that it was easy, once the reality of quanta was assumed, to explain a phenomenon called the photoelectric effect. Certain metals had been found to generate electric currents when light struck them, but physicists who had modeled the light strictly according to Maxwell, as a stream of wave energy striking the metal like ocean waves crashing into a shoreline, had been unable to explain the strength and other characteristics of the current produced. Einstein, picturing light instead as a bombardment of quanta, was able in a few simple lines of algebra to come up with the needed explanation. It was this achievement for which he won the Nobel Prize in 1921, perhaps on the grounds that of all his great works, this was the one that could most nearly be called a practical exercise in physics, in accord with Alfred Nobel's stipulation that his prize should honor accomplishments that carried with them technological benefits.

In his first paper of 1905, Einstein thus provided a theoretical and a practical reason to believe the quanta of radiation were real. This achievement made Boltzmann a grandfather to quantum theory on both sides, so to speak. Planck had first used his methods to derive the radiation formula, and now Einstein had used a different element of Boltzmann's work to show what that formula really meant.

The second paper, coming in May of the same year, bore even more directly on Boltzmann's view of physics. In it, Einstein provided an elegant explanation for a puzzling observation that went by the name of Brownian motion. In 1828, the Scottish botanist Robert Brown described how pollen grains, observed through a microscope, moved irregularly and incessantly, jiggling around as if they had some tiny life of their own. The origin of this constant agitation puzzled scientists for almost a century; some suggested it proved the existence of a kind of "life force" in organic entities, for which no explanation in inorganic physics could be possible.

Einstein's explanation was striking in its simplicity: the pollen grains were being buffeted this way and that by atoms in the air or liquid around them. A simple explanation, but not such an easy one to back up with physical argument. A single atom bashing into a huge, lumbering pollen grain can have no more effect than a pollen grain falling on a leaf. But, Einstein observed, there are many atoms bashing into a pollen grain at any one time, and they bash incessantly. Generally speaking, the more atoms collide with a grain, the more the effects of the impacts tend to cancel each other out, since they are moving essentially at random. But even in random motions there are chance fluctuations. It may just happen that a lot more atoms will hit one side of a pollen grain than another at a particular moment. This was the physics Einstein perceived and found a way to calculate. Using the standard Maxwell-Boltzmann formula for the distribution of atomic velocities, Einstein worked out the frequency and magnitude of collective collisions big enough to nudge a pollen grain off its track. His calculations amounted to a way of inferring the size and number of atoms from the direct mechanical effect of their impacts.

Einstein's explanation of Brownian motion stands as a turning point in the long debate over the existence of atoms. This was not only an observable and calculable effect of atomic motions, but a truly microscopic one. There was, moreover, no other plausible explanation for Brownian motion. Mach himself had always stressed the primacy of fundamental observations and directly detectable phenomena. Brownian motion was surely just such a phenomenon, and Einstein had now provided an atomic explanation for it. Three years later, in 1908, the French physicist Jean Perrin conducted a series of careful experiments that showed the accuracy of Einstein's formula for Brownian motion under a range of condition. For many doubters, Einstein's analysis, especially when it was backed up by Perrin's measurements, constituted the first empirical demonstration that atoms were real–tiny, hard, mobile objects, moving in predictable ways according to Newton's laws.

These two works of Einstein's became known in the first half of 1905, before Boltzmann had set off for his summer in California. Between them, they demonstrated the utility of Boltzmann's statistical methods in a new arena and provided an almost tangible proof of the existence of atoms. And yet there is no evidence that Boltzmann became aware of what Einstein had done. Early in 1905, he was mired in his abortive efforts to produce a monograph on his philosophical thoughts, and if his correspondence with Brentano gives an accurate indication, most of his intellectual effort was devoted to that project. Then he was in America for the summer, traveling around and griping about the food and drink. By September, when he was back in Vienna, word of Einstein's work must surely have been in the air. By then Einstein had also produced the third and fourth of his papers of that year, which contained what is now called the special theory of relativity. The last of these contained the famous formula $E = mc^2$. Still, none of this seems to have impinged on Boltzmann's consciousness.

It is hard to do away with the benefit of hindsight in trying to understand what happened in 1905. No one called it an *annus mirabilis* at the time. Both relativity and quantum theory puzzled many physicists for a long time. Planck himself, it appears, did not truly embrace the idea of quanta for almost another five years. There were many physicists who in 1905 saw Einstein's work, if they noticed it at all, as theoretical speculation of the highest degree, far removed from their immediate concerns. Many others must have looked at it and failed to understand its import.

Boltzmann, moreover, had his own reasons for not being interested in the first of Einstein's papers, on quanta. This was an elaboration or extension of Planck's work, and he had never had much time for Planck.

Much more puzzling, even astonishing, is Boltzmann's lack of awareness of Einstein's analysis of Brownian motion. Ten years earlier, in his reply to Zermelo, he had hinted at the very point that Einstein so triumphantly elaborated. Zermelo, Boltzmann had said, was trying to argue that kinetic theory was wrong because it predicted that the second law of thermodynamics was not absolute, that occasionally changes could happen that momentarily decreased entropy. Boltzmann's response was clear. Chance events of this sort are not a blemish but an inescapable element of the theory, and a testament to its validity. And he gave an example: "the observed motions of very small particles in a gas may be due to the circumstance that the pressure exerted on their surfaces by the gas is sometimes a little greater, sometimes a little smaller." In other words, tiny variations in pressure due to the irregular impact of atoms or molecules cause very small particles to jiggle about. Boltzmann quite clearly understood, in 1896, that Brownian motion is an elementary consequence of kinetic theory.

But he threw the point out as an aside, in the course of an abstruse mathematical correspondence over the niceties of kinetic theory which, perhaps, few other physicists cared to read. Nor did Boltzmann attempt to quantify his notion, to see if atomic motions in a gas indeed produced fluctuations of the right scale to explain Brownian motion. He was surely capable of such a calculation, but apparently didn't grasp how persuasive the conclusion might be. If, as it appeared to Boltzmann at the time, few of his fellow physicists had any regard for kinetic theory, why should they pay attention to a kinetic explanation of Brownian motion?

There may also have been a partly psychological reason behind Boltzmann's failure to pursue the point. In battling Zermelo he found himself saying again, as he had said so many times before, that tiny abrogations of the second law of thermodynamics were rare and fundamentally unimportant. That they happened was undeniable; that they mattered was less clear. The subject was distasteful and wearisome to him by now. He regarded the occurrence of fluctuations as a straightforward manifestation of kinetic theory, not a flaw, and if Zermelo and the rest couldn't see that, it was hardly worth the effort of trying to convince them. So he took the matter no further.

The fact remains, however, that Boltzmann understood in 1896 the essence of what Einstein became famous for explaining in 1905. One can hardly imagine how different Boltzmann's last years might have been if he had taken the trouble to publish even a rough calculation of Brownian motion, instead of letting the idea slip out as an incidental suggestion.

Then again, was the time ripe? Were physicists in 1905 somehow more receptive to the idea than they might have been 10 years earlier? Half a century before, John Waterston had written an outline of kinetic theory that fell on deaf ears, while Rudolf Clausius, offering a much simpler exposition just 12 years later, won the world's attention. Sometimes scientific ideas, like strange musical compositions or surrealistic dramas, need a ready audience as well as a creator.

Sadder still, the period from 1896 to 1905, encompassing the disillusionment of his return to Vienna, the desperate escape to Leipzig, and his final, resigned homecoming, saw Boltzmann himself become part of the uncomprehending audience to the new physics that was emerging. His inattention to Einstein's revelatory calculation, when it finally did appear, may have derived from entirely mundane reasons. He was busy with other things, and his health and state of mind continued to be fragile. Immediately after returning from California, he had plunged into work. He was supposed to contribute a long article to a comprehensive Mathematical Encyclopedia, edited by his friend Felix Klein, on kinetic theory–a task he had agreed to some time before but had been continually putting off. Klein knew that Boltzmann was the right man for the job, perhaps indeed the only person who could do it at that time with sufficient conviction.

Anxious to get Boltzmann's essay for his encyclopedia, Klein had resorted to a joking threat: if you don't do it, he said, I will get the essay from Zermelo instead. Boltzmann had at last been stung into action. "Before that Pestalutz does it, I will do it," he told Klein. Pestalutz is an exceedingly minor nonspeaking character in Schiller's play *The Death of Wallenstein*. Late in the drama, conspirators are haggling over who is to assassinate Count Wallenstein; "I wouldn't leave it up to Pestalutz," one of them says. Pestalutz is then himself dispatched, as a precautionary measure, a turn of events that may have cheered Boltzmann when he thought of Zermelo.

As soon as he arrived in the north German port of Bremen on the ship from New York, Boltzmann had written to his assistant in Vienna, Stefan Meyer, asking him to find someone to help him with reading and research for the article. For some years now he had relied on young physics students or sometimes his wife or son to read aloud to him, his eyesight having failed almost completely.

At the end of the year, the job was finally done. He also worked with a publisher to produce a volume of his more general essays, which came out at the end of 1905 under the title *Populäre Schriften,* or *Popular Writings*. Many of these were reprinted versions of articles and lectures he had delivered in years past, but he also managed to include his account of the California trip. Between writing that and finishing his encyclopedia entry, Boltzmann can have had little spare time in the last quarter of 1905. When all was done, he relapsed into depression and illness. He was bedridden through Christmas and the New Year, and wrote to Brentano of how his spirit remained low even after the physical illness had dissipated. "How I envy you your constant cheerfulness and satisfaction. You are in truth a genuine philosopher. I have reached 62 years of age and I have gained no peace of mind." A week later, writing to his old friend Arrhenius, he said, "Unfortunately, things are not going especially well with me. I am suffering fearfully from my old evil, neurasthenia. I definitely overexerted myself in California."

In February he let Meyer know that he was no longer capable of lecturing and arranged to send over all his student records. He decided at some point that the little cottage that he had so much enjoyed just three years earlier was now "unhygienic," and he set up makeshift living arrangements in the physics building on Türkenstrasse. His wife and daughters evidently remained in the cottage. Sometime early in 1906, he managed to participate in a final oral exam for one of his students, Ludwig Flamm, who later married his youngest daughter. Flamm reported that on his way out, he heard "heart-rending groans." Another physicist who visited Boltzmann around Easter of that year heard him lament out loud, "I would not have believed I could come to such an end."

By May 1906, university officials acknowledged that Boltzmann could no longer teach. Meyer was taking care of physics. As for the sporadic philosophy class that would now go untaught, it was tartly noted that "no pressing pedagogical need will be damaged by that." Thus, over the months following his return from California, did the physicist and sometime philosopher disengage himself from the intellectual concerns that had been the mainspring of his life.

On the morning of Friday, September 7, 1906, a reporter for the Vienna *Neue Freie Presse* made the 20-minute journey from Duino, a resort village on the Adriatic coast not far from Trieste, to a small nearby chapel. Along the way he encountered a "woeful sight." Two wagons were toiling across the picturesque but sparse landscape, bearing an obviously distraught company of people. The first carried a woman of late middle age and her three daughters, the second a young man in military uniform and a priest. The carts clattered on to Trieste, to meet the train for Vienna. The reporter pressed on to the small chapel of San Giovanni, where he found the black-draped body of a short, heavyset man set out on a bier. This was the man whose absence from the company in the two carts was the source of its misery: the husband of the woman in the cart, the father of the three young women and the junior officer, and the man for whom the priest had last night said a requiem mass in the chapel. There had been some delay in having the body removed to the chapel and prepared for the mass. A commission of local justice officers had been called from Trieste to investigate the circumstances of the death, and before making an official declaration of what was obvious to all–the man had been found by one of his daughters, hanging from the window frame in his hotel room–the justice commissioners had to interview the hotel keeper and his staff, the shocked and grieving family, the doctor who had been hurriedly summoned to pronounce that the man was indeed dead. Only then could the body be removed to the peace and quiet of the chapel of San Giovanni. While the reporter gathered together these details of the circumstances surrounding this suicide, the body was made ready for transport on the evening express to Vienna. There, in the Friday evening edition of the *Neue Freie Presse,* the tragic tale was recounted.

The newspaper offered much more than a plain account of a solitary death in a distant village. The suicide was Ludwig Eduard Boltzmann, a native son of Vienna, 62 years old, and a physicist and mathematician of international renown. He was a man who "bestrode his time and his nation," but his ill health and neurasthenia had been well known to friends and colleagues.

Boltzmann, with his wife and daughters, had traveled to the coastal village of Duino about three weeks earlier to recover himself before the beginning of the new academic year. The respite by the sea had seemed to calm Boltzmann's mood. He had gone on rambles in the countryside and sent postcards back to friends and colleagues in Vienna. Early in September, much of the family's baggage was sent ahead to Vienna, but the imminence of the return abruptly changed Boltzmann's mood. Concerned, his wife telegraphed their son, who had just got back to Vienna after participating in military exercises as a volunteer officer, and told him to come to Duino immediately.

About 6 o'clock on the evening of Wednesday, September 5, Henriette Boltzmann and her daughters went down to the sea to bathe, leaving her husband behind in his room at the hotel. He said he would join his family later, but the evening wore on and he did not come. Henriette became worried, but apparently not all that worried, since she sent her 15-year-old daughter, Elsa, back to the hotel to fetch her father. It was she who with "unspeakable horror" found him hanged from the sturdy crossbar of a window. For the rest of her life, she never spoke of her terrible discovery. Arthur Boltzmann, the son, arrived too late to see his father alive.

News of Boltzmann's death threw the University of Vienna into "active mourning," the *Neue Freie Presse* declared. As well as the reporter's account from Duino, the newspaper carried a description of the scene at the dead man's Vienna household, and printed two eulogies from colleagues and admirers. Next morning's paper carried two more.

Among the dignified, decorous, occasionally rapturous remarks of his obituarists were frank admissions that Boltzmann's mental health was not of the strongest. His struggles with neurasthenia were mentioned. His colleague Franz Exner sketched Boltzmann's scientific achievements, his mathematical prowess, early success and far-reaching reputation, and asked rhetorically whether so gifted a man must not inevitably be happy. No, Exner answered himself, he was not: as if to counterbalance Boltzmann's prodigious talents and his far-seeing intellect, "envious fortune denied him inner peace."

Ernst Mach, too, weighed in with an assessment of Boltzmann's life and achievements, and suggested that his "delicate, sensitive nerves" were ill-suited for the hard grind of the laboratory and the lecture hall. At the beginning of the 20th century, Mach was already fretting that the proliferation of scientists and scientific journals had made the world of research an excessively competitive place. No mention of the fact that it was Mach's own opposition that had inspired at least some of the controversy that had been so burdensome to Boltzmann. Indeed, Mach airily observed, the surprising thing was perhaps that cases such as the unfortunate death of Boltzmann did not occur more often.

In Sunday's newspaper, Henriette Boltzmann posted an announcement thanking friends, colleagues, and students of her husband for their commiserations and warm wishes. At Boltzmann's funeral a few days later, only two physicists, Gustav Jäger and Stefan Meyer, attended. Both spoke of their loss. It was the end of summer vacation, and the university had not yet sprung into life for the coming academic year.

IN THE YEARS following Boltzmann's death, physics changed utterly. Any lingering debate over the reality of atoms soon dissolved. Wilhelm Ostwald, Boltzmann's lifelong friend and antagonist, was too good a scientist and too honorable a man to hold on to his anti-atomic views for very much longer. As early as 1908, when he was writing a new introduction to his standard textbook *Outline of General Chemistry,* he overtly stated his belief in the existence of atoms. He did not entirely abandon his old philosophical fondness for the cause of energetics, or at least for seeing energy as somehow the fundamental element of natural science, but the now-proven reality of atoms along with the long-standing difficulties in turning the idea of energetics into any kind of useful, quantitative theory combined to deflate energeticism as a worthwhile cause. It simply faded from scientific discussion.

Mach was a harder case. His philosophy continued to have adherents, although fewer within science and more outside it. Planck fiercely denounced Mach as a "false prophet" in 1908. Einstein, who had once been enamored of Mach's puritanical dictates on the right way to construct physical theories, came to see the poverty of his viewpoint, and turned against his teachings. Special relativity owed something to Mach's strictures; it was by strict attention to observational detail, by thinking carefully and precisely how an observer would seek to establish or verify the simultaneity of events happening in distant places, that Einstein understood why time is "relative"–why it is not the same for everyone. In line with Mach's prescription, Einstein saw that time means what an observer measures it to mean.

Yet Mach disliked and argued against relativity–it was another theory, after all–and Einstein, as he delved further into the implications of his new ideas, saw that it was essential to go beyond mere observation. It takes theories to build the factual bricks of the scientific world into coherent structures. Speaking of another physicist, who was still an enthusiast for Mach, Einstein remarked that he "rides Mach's poor horse to exhaustion," and that this horse "cannot give birth to anything living, it can only exterminate vermin." Later still, Einstein was bluntly dismissive: "Mach was a good experimental physicist but a miserable philosopher"; he made "a catalog not a system." As late as 1915, Mach was still writing in an anti-atomic vein. He died the following year, at the age of 78.

A few short years after Boltzmann's death, his physics had been vindicated. The existence of atoms was unquestioned; kinetic theory was unarguable. On the question that had dogged Boltzmann for so many years, over the probabilistic nature of the second law of thermodynamics, the simple conclusion was that he was right and his critics were wrong. The laws of thermodynamics are, in fact, approximate laws: heat almost always flows from hot to cold; entropy almost always increases. It may have been a shock to several classical physicists that the laws were not absolute, but succeeding generations of physicists have got over the shock.

Still, by the time of his death, Boltzmann's writings on the subject of reversibility contained, on close examination, a variety of incompatible arguments. Sometimes the H-theorem was treated as implying an absolute increase in entropy; elsewhere he acknowledged that the tendency of entropy to always grow was not inevitable but only very likely.

Clarity was in the end brought to much of this work by one of

Boltzmann's more gifted students, Paul Ehrenfest. He delved into all the kinetic and probabilistic and statistical arguments and, with his Russian-born wife Tatiana, also a physicist, published a thorough and rigorous analysis in 1911 in the form of a lengthy article for Felix Klein's mathematical encyclopedia. The Ehrenfests thought the whole subject through from beginning to end and set the results down all at once, in a fully consistent way–an achievement that Boltzmann himself was perhaps not truly capable of. The Ehrenfests' work did much to clarify the true import of Boltzmann's original ideas and methods.

The source of Boltzmann's confusion over the meaning of his H-theorem was, it eventually became clear, a somewhat concealed technicality of his derivation. In deriving it, he had made the pioneering step of analyzing the effects of collisions between atoms on the distribution of atomic velocities, but in order to render the problem tractable, he had been obliged to make some assumptions about the characteristics of typical or average collisions. However, it is precisely the atypical or non-average atomic collisions that cause H to increase and entropy to decrease, in contradiction to the second law. Boltzmann's proof of the H-theorem thus embodied a subtle assumption that had the effect of ruling out the odd and unusual atomic motions that would disobey the theorem. Boltzmann's acknowledgment that in fact H could, sometimes, increase instead of decreasing was physically the correct conclusion, but he never looked back at his own work to see how this admission was to be reconciled with his own mathematics.

The dispute with Zermelo had repercussions that continued to reverberate for many years. Poincaré's theorem, on which Zermelo based his objection to kinetic theory, asserted that any dynamical system, such as a collection of atoms, was guaranteed to return to its starting point sooner or later. In proving this result, Poincaré made use of a new perspective on dynamics that was to become widely influential. Much as Maxwell and Boltzmann had themselves envisaged a gas jumping endlessly from one possible state to another, as its atoms collided, so Poincaré in a more general sense depicted each possible state of a dynamical system as a point in an abstract mathematical space. As the system evolved according to standard mechanical laws, it moved from one conformation to another, tracing out a path or trajectory in this space. His theorem is then a proof that this trajectory is a closed loop, so that given enough time, the system will traverse the whole path and return to where it began.

This point of view allows the mathematician to represent the behavior of even the most complex system as a trajectory through a suitably defined space. The modern theories of chaos and complexity depend on exactly this kind of analysis. A chaotic system is one in which trajectories in this dynamical space that start out very close to each other diverge with exponential rapidity, so that two systems that begin in almost the same state very quickly end up far apart. So rapid is this divergence that the system becomes in effect unpredictable, or chaotic. As a practical matter, it is impossible to specify starting points with enough precision that their subsequent evolution can be accurately predicted.

This understanding, only recently achieved, provides another rationalization of the apparent contradiction between Poincaré's recurrence theorem and Boltzmann's H-theorem. In a system with a realistically huge number of atoms, predicting the exact dynamical evolution becomes, practically speaking, impossible. Even though such a system is fully deterministic, in the sense that every state turns through exact mechanical equations into another state, its behavior is to all intents and purposes unpredictable, or random, because the states cannot be specified with sufficient accuracy.

This, therefore, is one answer to the puzzle that both Maxwell and Boltzmann came up against. They assumed, but could not prove, that a system of atoms would behave in an essentially random way, even though the underlying mechanics laws are fully deterministic. Maxwell balked at this difficulty more than Boltzmann did, which is one reason the Scotsman never pursued kinetic theory as far as he might have. Boltzmann, by contrast, prized his sense of the physical rightness of his thinking above its strict mathematical consistency, and his faith, much later, proved well founded. This is another illustration of the value of Boltzmann's philosophical attitude: in any theorizing there will always be unproven assumptions or technical problems that can't immediately be solved, but the important thing is to keep pushing forward. Insisting that every step be absolutely irrefutable is a recipe for going nowhere fast. Boltzmann might well have borrowed Emerson's phrase: a foolish consistency is the hobgoblin of little minds.

BOLTZMANN LEFT a legacy of scientific achievement that laid the groundwork for quantum theory and held clues to chaotic dynamics, but for many years before he died he had been caught up in a morass of philosophizing. He had lost touch with physics and was struggling instead to achieve realizations or certainties in a subject which he did not fundamentally believe to have the capacity for such things.

In his last year the despair became overt. "Nothing more will come from me," he confided to Meyer. Physics had been the mainstay of his life, but he no longer kept up with it. He had delivered his own summation of his life's work in the *Lectures on Gas Theory*, and in the introduction to the second volume, he made it abundantly clear that he was writing for the future. He had given up hope that his own contemporaries would abandon their hostility and read his work with open minds. It was for a new generation of physicists, unencumbered by the philosophical burdens of the old, that he was setting his work down. His achievements in physics would be in the libraries forever, awaiting fresh attention when the time was ripe. His brief attempt to engage the philosophers of his own generation on their own ground had proved futile. This was no worthwhile arena for a man of his intellect. And now, at last, his eyesight was so poor he could no longer read for himself. It was even a struggle to play the piano.

No one who lived in Vienna around the turn of the century can have been a stranger to the thought of suicide. There was almost a mania for it. In 1889, Franz-Josef's son, the crown prince Rudolf, shot his teenage mistress and then killed himself. That left no direct heir to the Habsburg throne, and Franz-Josef's nephew Franz Ferdinand begin to draw power around his circle of political allies and hangers-on. He established himself in the eastern part of the empire, where he was for the most part cordially disliked. His assassination in Sarajevo in 1914 triggered the First World War, which, among other things, destroyed Austria-Hungary as a political entity and the Habsburgs as a dynasty.

Suicide was pervasive in Vienna. A tightrope walker tied a rope around his neck, attached the other end to a window frame, and jumped. A circus acrobat argued with his wife just before going on, and jumped spectacularly from the high trapeze. Ernst Mach's son died by an overdose a few days after receiving his doctoral degree. An architect of the Opera-House on the new Ringstrasse in Vienna killed himself, it was said, after the emperor expressed some reservations about the building; it was after this incident that Franz-Josef hit on the cautious formula "it was very nice; it pleased me very much" when confronted with any request for an aesthetic evaluation. One of Franz-Josef's finance ministers had killed himself when the economy took a downturn. Especially toward the century's end, Viennese both low and high threw themselves into rivers, off trains, and out of windows in alarming numbers. Boltzmann's student Paul Ehrenfest, who did so much to establish his teacher's place in science, killed himself in 1933 at the age of 53.

Even though the backbone of the Viennese way of life was *fortwürsteln* (muddling through), there comes a time when circumstances are so hard that muddling through doesn't work any more. Then one has to fall back on reserves of inner strength, on principles and beliefs. And if there are no such principles or beliefs, perhaps death begins to seem the only option.

None of this can explain why any one individual would choose suicide, but Boltzmann lived in a city and at a time in which killing oneself was tinged with both romanticism and dignity. His friends and colleagues had worried for years that he might, in a low mood, take his own life. His low moods had become increasingly frequent, and increasingly severe. Throughout his life, Boltzmann had embraced controversy, striking back sharply at his enemies even, in some cases, when his enemies were also his friends, but these attacks damaged him as much as they hurt anyone else. He lacked a certain robustness of spirit; he took criticism personally; he couldn't bear for erroneous arguments to go unanswered. Mach, one may imagine, derived a good measure of self-satisfaction from knowing that his views caused consternation, irritation, and dispute. He had been disputatious as a schoolboy; it proved that he had something worth saying. Boltzmann, although he may in his own heart have had confidence that he was right, needed public affirmation of his rightness. Ostwald, in his autobiography, speculated along these lines. It was Boltzmann's very earnestness in needing to deal with every criticism and contradiction that made him excessively sensitive. He could not distinguish between knife wounds and fleabites. Lise Meitner made the same point: "He may have been wounded by many things a more robust person would hardly have noticed. . . . I believe he was such a powerful teacher just because of his uncommon humanity."

A hint of Boltzmann's torn nature can be seen in the obituary notices he wrote for Stefan and Loschmidt. He rebuked them a little for their failure to travel, to go beyond the confines of Vienna, to promote themselves among their colleagues: "those who strive equally for great accomplishment as well as external recognition appear as the more complete, the better." And he recalled an English colleague who thought to compliment him by observing, in the context of Austria's loss of prestige and territory to the Prussians, that the Austrians were "too good" to win. "We shall have to rid ourselves of this goodness and self-contentment," Boltzmann sharply observed.

He cited Loschmidt as a regrettable example of the typical Austrian, reluctant both to praise and to be praised, and he wondered where this odd characteristic comes from: "Is it that those who easily overcome the greatest difficulties in analytical thinking and experiment face difficulty in exhibiting the skill of valuing oneself, which comes so easily to many people of no great significance, or is it rather the greatest quality of the soul to be indifferent to external recognition? I do not know." The thought that Stefan and Loschmidt, through a modest recognition of their own worth and a disregard for plaudits from others, might have achieved a certain quiet contentedness is one that seems not to have broached Boltzmann's consciousness. His own life hardly amounts to a warm recommendation for the strategy of endless travel and insistence on seeing one's work properly esteemed.

Among his other intellectual enthusiasms, Boltzmann became a great promoter of the new ideas of Darwin. He saw early on that life itself was a thermodynamic phenomenon, since living creatures are physical engines that must draw energy from the world around them. But he also perceived a subtlety to that argument: "The overall struggle for existence of living beings is therefore not a struggle for raw materials–the raw materials of all organisms are available in excess in the air, water, and ground–nor for energy, which in the form of heat is plentiful in every body, but rather a struggle for entropy, which becomes available in the flow of energy from the hot sun to the cold earth."

The realization that the maintenance of life demanded a constantly productive thermodynamic interaction with the environment gave him an immediate sympathy for Darwin's ideas. Evolution was, so to speak, a kind of statistical mechanics. As individual creatures interacted, lived and died, prospered or suffered, so the overall qualities of a species emerged. Through the application of a simple rule–survival of the fittest–evolution had the power to explain the complexity of life in all its manifestations.

He imagined that Darwinism could apply not just to the physical form of creatures, but to their intellectual capacities and even to ethical feelings or spiritual desires. The ability to understand the world represented an evolutionary advantage. In his inaugural lecture on becoming a professor in Leipzig, Boltzmann had elaborated on this theme:

"We see how it was useful and important for our species that certain sense impressions were flattering and therefore sought after, while others repelled us; we see how advantageous it was to construct in our minds the most accurate pictures possible of our surroundings and strictly to keep apart the true ones, those which correspond with experience, from the false, which do not. We can therefore explain the genesis of an apprehension of beauty as well as of truth in terms of mechanics."

Truth is beauty, in other words, and truth has survival value. Therefore evolution favors the development of an appreciation of beauty. Boltzmann became fond of such "mechanical" explanations of aesthetic or moral qualities in humans. The sorts of explanations he thus produced may have been rather glib and superficial, but the general idea has become popular again in recent years, as evolutionary biologists have sought to understand how human behavior, emotions, psychology, and so on, spring from the Darwinian struggle.

Boltzmann also recognized, however, that his argument is a two-edged sword: if a certain quality promotes survival, it follows that the absence of that quality can be fatal:

"We understand why only those individuals could continue to exist which abhorred certain very noxious influences with all the nervous energy at their command . . . while with equal vigor aiming at other influences that were important for the preservation of themselves or their kind. In this way we grasp how the intensity and power of our whole affective life developed: pleasure and pain, hate and love, happiness and despair. We cannot rid ourselves of the whole range of our passions any more than we can our bodily illnesses, but on the other hand we learn how to understand and bear them."

Boltzmann spoke these words in 1900. A few years later he came to the end of his own capacity to understand and bear the passions of his life.

Postscript

THE YEAR 1900, WHEN MAX PLANCK brought quantum theory into the world, serves a little too conveniently as the pivot of a scientific revolution. Planck's solution of the radiation problem had its roots in Boltzmann's earlier studies of kinetic theory, and the true significance of what he had done did not become plain until some years into the new century. Even so, there was indeed a revolution. The 20th century became the century of modern physics, from quantum theory and relativity to nuclear and particle physics and big bang cosmology. Preceding it, stretching back into the past, was the calmer, more certain era of classical physics, of heat, light, and sound, of mechanics and thermodynamics.

Or so the history of physics is often portrayed. From the vantage point of the early 21st century, this thumbnail sketch seems ludicrously oversimplified. Modern physics is now a century old, which means it has lasted just about twice as long as the golden age of classical physics–if ever there was such a period. In 1847, a few years after Ludwig Boltzmann's birth, Helmholtz published his thorough enunciation of that most classical law, the principle of energy conservation. The second law of thermodynamics came later, from Clausius. Then there was Maxwell's electromagnetism, and his and Boltzmann's sophisticated study of kinetic theory and the foundations of statistical mechanics, which Gibbs completed in 1903. Apart from Newtonian mechanics, the fundamental elements of classical physics appeared within little more than a 50-year stretch, covering the second half of the 19th century.

And never, within that short span of time, was there a period of stability and consensus. Arguments over kinetic theory and its statistical connotations, and over the nature of Maxwell's electromagnetic field, persisted into the beginning of the 20th century–after quantum theory had put in its fledging appearance. Boltzmann's lifespan conveniently coincides with the age of classical physics, which was a time of constant debate, evolution, and controversy.

Nor did the transition from classical to modern physics mark such a wholesale conceptual change as first appears. Quantum mechanics, especially with Werner Heisenberg's formulation of the uncertainty principle in 1926, implies a more deeply probabilistic view of nature than kinetic theory, but the earlier debate over the second law of thermodynamics had loosened the soil in which this later branch of statistical thinking was to grow.

But most profoundly of all, the modern idea of theoretical physics had, by virtue of Boltzmann's struggles, already established itself by the time quantum theory and relativity came along. During the 20th century, physicists dreamed up particles such as neutrinos, positrons, and quarks, and only later obtained experimental evidence for them. Such proposals were sometimes controversial, but nobody complained that they were intellectually invalid, as Mach had argued of Boltzmann's atoms. Today, cosmologists debate what the universe looked like when it was much less than a trillionth of a second old and far tinier than an atom. Theorizing of this sort may be extravagant, but it is not generally reckoned unscientific.

This, in the largest sense, is the legacy of Boltzmann's difficult victory over Mach. To some extent, the debate continues today. Do physicists and cosmologists who explore the idea of a universe built from superstrings wriggling around in eleven dimensions sometimes stray into theorizing so gloriously speculative that it cannot possibly be tested? Perhaps. But the debate is not about whether theorizing of this sort is philosophically legitimate, but whether it is useful, whether it can bring enlightenment rather than confusion. The right to theorize in this way is what Boltzmann fought for throughout his life, in the face of derision from his critics. Judge theories by what they do, by the kind of further thinking they engender, not by their conformity to some excessively simple-minded version of common sense.

Mach's influence continued for a while in the philosophical movement known as the Vienna Circle, whose members espoused a brand of thinking called logical positivism. Their aim, to grossly oversimplify, was to begin with unquestionable empirical facts and use prescribed lines of reasoning to construct acceptable scientific ideas. But even in the utterly logical realm of mathematics, the rigidities of positivism proved excessive. First Bertrand Russell and then Kurt Gödel showed that within any finite mathematical system, there are always objects that can't be unequivocally classified and theorems that can't be proved right or wrong.

In the end it was Boltzmann's style of thinking about science that proved more fruitful. He was never able to formulate a philosophy of science in a perfectly consistent and logically complete way, but that failure is now seen as a consequence of the necessarily creative and adaptable nature of science. No formula for scientific progress can be devised. No one tries any more to think up universal philosophies of science.

Boltzmann's philosophy of sophisticated pragmatism, tentative and inarticulate thought it may have been, found continuation in the work of two important 20th-century thinkers, both Viennese. Karl Popper observed that theories can be proved wrong but never right, and he argued that a theory acquires credibility, and comes to be regarded as a good theory, by passing increasingly rigorous tests. Abstract definitions of truth or reality have no part in this process. Popper's depiction is close to Boltzmann's view of theories as ever-closer approximations to some possibly unreachable final answer. To the extent that scientists see any need for a philosophical justification of their endeavors, this sort of portrayal generally suffices.

Another idea implicit in Boltzmann's dispute with Mach was that theories bring about new understanding by creating new concepts, ones that were not apparent beforehand. It was only after Clausius had defined the new thing called entropy, for example, that the second law of thermodynamics could be discerned. The scientist must view the natural world through the lens of theory.

This notion attracted the interest of the young Ludwig Wittgenstein, who had hoped to study under Boltzmann. But Wittgenstein was just 17 when Boltzmann died, and he left Vienna to study mathematics and engineering. His later philosophical development nevertheless retained traces of Boltzmann's ideas. Wittgenstein argued that many supposedly philosophical questions were in fact the result of confusion over language and definitions, and that when terms were consistently defined, the questions either resolve themselves unambiguously or are revealed to be empty or self-contradictory. The dispute between Mach and Boltzmann over the existence of atoms could never be resolved by philosophy, because each was judging the worth of atoms against different standards. The point of philosophical inquiry, according to Wittgenstein, is to figure out whether questions make sense. As he famously put it, "whereof we cannot speak, thereof we must remain silent."

BOLTZMANN LEFT the stamp of his intellect on the very idea of modern theoretical physics. His personal imprint, on the other hand, was weak. He founded no school and gave only sporadic guidance to a few young researchers, notably Nernst and Arrhenius, who went on to great achievements of their own. Lise Meitner and Paul Ehrenfest were the last two physicists of wide repute to emerge from the declining Vienna of Boltzmann's final years. Erwin Schrödinger, destined to devise the wave equation for quantum mechanics that bears his name, enrolled as an undergraduate at the University of Vienna in the fall of 1906, eager to hear Boltzmann's lectures, but like Wittgenstein, he was too late.

Not long after Boltzmann's death, Vienna itself flew to pieces.

The mix of pragmatism, loyalty, and muddling through that had kept the Austro-Hungarian empire shakily intact finally exploded into the First World War. Franz-Josef died in 1916, by then a frail old man of 86. His nephew Karl became the new emperor, but the monarchy lasted only as long as the war, ending in 1918 when the nations of Austria-Hungary went their separate ways. Austria itself was left small and impoverished.

Henriette Boltzmann survived the upheavals and, perhaps happily for her, didn't quite live to see the next round. She died in 1938, at the age of 84.

During the chaos of the early 20th century, Boltzmann's grave fell into neglect. The plot had been paid for only for a period of 20 years, and some time later it was augmented with a new occupant. In 1929, through the efforts of scientists and others, Boltzmann's coffin was removed from its old resting place (a tricky operation, as the family of the person buried above would not allow their relative to be disturbed, and a shaft had to be dug at an angle to retrieve Boltzmann), and the physicist was buried at a new site in Vienna's Central Cemetery. In 1933, a monument was erected, adorned with a bust of Boltzmann, and inscribed with the deceptively simple formula that encapsulates one of the most profound achievements of the classical era of physics, a formula that also presaged the end of that era and the coming of a new age:

$$S = k \log W$$

Acknowledgments

I SPENT MANY A PLEASANT AND PRODUCTIVE DAY exploring the historical collections of the Niels Bohr Library of the American Institute of Physics in College Park, Maryland, and I am grateful to the staff there for their ready assistance. I also made considerable use of the libraries of the University of Maryland in College Park and the Library of Congress, and I am happy to acknowledge the services provided by these public institutions and their staffs.

Ralph Cahn in Munich assisted me by finding a number of German and Austrian books and documents, and I am grateful to Toni Feder and Wolfgang Frey for helping me with translations from German. Stephen Morrow, my editor at The Free Press, provided many illuminating comments and suggestions on earlier versions of the manuscript.

Bibliography and Notes

A FULL-SCALE BIOGRAPHY OF LUDWIG BOLTZMANN has yet to be written, nor is this book intended to make up that omission. Details of his life, especially of his early life, are scarce and come mostly from the recollections and anecdotes of those who knew him, in some cases not particularly well. Boltzmann never recorded anything approximating a memoir, and his more informal published writings refer only rarely, and in passing, to incidents in his own life.

A sketch of Boltzmann's life forms the first section of Engelbert Broda's short book *Ludwig Boltzmann,* originally published in German in 1955. Broda collected many references and sources that have been cited in later accounts. More recently the three-volume documentary compilation edited by Walter Höflechner, *Ludwig Boltzmann: Leben und Briefe (Life and Letters),* has appeared, and I have relied heavily on this as a definitive guide to the details of Boltzmann's life. Höflechner's volume 2 contains all known correspondence to and from Boltzmann, except for the personal correspondence with Henriette von Aigentler, which has been compiled and edited by Dieter Flamm, who is a grandson of Boltzmann's.

For biographical details of many of the lesser characters, I have used the excellent *Dictionary of Scientific Biography,* Charles C. Gillispie (ed.), Scribners', New York, 1970.

I owe a great debt to all the scholars and researchers who have untangled, documented, and analyzed the details of Boltzmann's life and work. In describing Boltzmann's physics and the physics of his times, I have leaned particularly on Thomas Kuhn's *Black-Body Theory and the Quantum Discontinuity,* which includes a staggeringly thorough analysis of the H-theorem and its ramifications. Stephen Brush's writings, especially his *Statistical Physics,* as well as his translations of some of Boltzmann's technical works, also helped me a great deal.

Bibliography

Barea, Ilsa, *Vienna: Legend and Reality,* Pimlico, London, 1993.

Blackmore, John, *Ernst Mach: His Life, Work, and Influence,* University of California Press, Berkeley, 1972.

Blackmore, John (ed.), *Ludwig Boltzmann: His Later Life and Philosophy, 1900–1906,* Kluwer Academic Publishers, Dordrecht/Boston/London, 1995.

Boltzmann, Ludwig, *Lectures on Gas Theory,* University of California Press, Berkeley, 1964. Reprinted by Dover, New York, 1995 (translation by Stephen G. Brush of *Vorlesungen über Gastheorie,* J. A. Barth, Leipzig, 1896 and 1898).

Boltzmann, Ludwig, *Populäre Schriften,* J. A. Barth, Leipzig, 1905.

Broda, Engelbert, *Ludwig Boltzmann: Man-Physicist-Philosopher,* Ox Bow Press, Woodbridge, Conn., 1983 (translation of *Ludwig Boltzmann: Mensch, Physiker, Philosoph,* Deuticke, Vienna, 1955).

Brown, Sanford C., *Count Rumford: Physicist Extraordinary,* Doubleday, New York, 1962.

Browne, Janet, *Charles Darwin: Voyaging,* Princeton University Press, Princeton, N.J., 1995.

Brush, Stephen G., *Statistical Physics and the Atomic Theory of Matter,* Princeton University Press, Princeton, N.J., 1983.

Bumstead, H. A., and R. G. van Name (eds.), *The Collected Works of Josiah Willard Gibbs,* Longmans, New York, 1928.

Cahan, David, *An Institute for an Empire,* Cambridge University Press, New York, 1989.

Campbell, Lewis, and William Garnett, *The Life of James Clerk Maxwell* (2nd ed.), Macmillan, London, 1884.

Cercignani, Carlo, *Ludwig Boltzmann: The Man Who Trusted Atoms,* Oxford University Press, New York, 1998.

Cohen, E. G. D., and W. Thirring (eds.), *The Boltzmann Equation: Theory and Application,* Springer-Verlag, Vienna, 1973.

Crankshaw, Edward, *The Fall of the House of Habsburg,* Viking, New York, 1963.

Fasol-Boltzmann, Ilse (ed.), *Principien der Naturfilosofi,* Springer-Verlag, Berlin, 1990.

Flamm, Dieter (ed.), *Hochgeehrter Herr Professor! Innig geliebter Louis! Ludwig Boltzmann, Henriette von Aigentler Briefwechsel,* Böhlau-Verlag, Vienna, 1995.

Hasenöhrl, Fritz (ed.), *Wissenschaftliche Abhandlung von Ludwig Boltzmann* (3 vols.), Chelsea Publishing, New York, 1968 (originally published by J. A. Barth, Leipzig, 1909).

Hawking, Stephen, *A Brief History of Time,* Bantam, New York, 1988.

Hoffman, Paul, *The Viennese: Splendor, Twilight and Exile,* Doubleday, New York, 1988.

Höflechner, Walter (ed.), *Ludwig Boltzmann: Leben und Briefe,* Akademische Druck- und Verlagsanstalt, Graz, Austria, 1994.

Hörz, Herbert, and Andreas Laass, *Ludwig Boltzmanns Wege nach Berlin,* Akademie-Verlag, Berlin (DDR), 1989.

Janik, Allan, and Stephen Toulmin, *Wittgenstein's Vienna,* Simon & Schuster, New York, 1973.

Knott, C. G., *Life and Work of Peter Guthrie Tait,* Cambridge University Press, New York, 1911.

Kuhn, Thomas S., *Black-Body Theory and the Quantum Discontinuity, 1894–1912,* University of Chicago Press, Chicago, 1978.

Lucretius, *De Rerum Natura,* edited and translated by Anthony M. Esolen, Johns Hopkins University Press, Baltimore, 1995.

Malcolm, Norman, *Ludwig Wittgenstein: A Memoir,* Oxford University Press, New York, 1958.

McGuinness, Brian (ed.), *Theoretical Physics and Philosophical Problems,* D. Reidel, Dordrecht/Boston, 1974 (translation of a selection of Boltzmann's writings, including some but not all of the *Populäre Schriften*).

Millikan, Robert A., *The Autobiography of Robert A. Millikan,* Arno Press, New York, 1980.

Moore, Water, *Schrödinger: Life and Thought,* Cambridge University Press, New York, 1989.

Morton, Frederic, *A Nervous Splendor: Vienna 1888/1889,* Penguin, New York, 1980.

Niven, W. D., *The Scientific Papers of James Clerk Maxwell* (2 vols.), Cambridge University Press, Cambridge, 1890.

Ostwald, W., *Grosse Männer,* Akademische Verlagsgesellschaft, Leipzig, 1909.

Palmer, Alan, *Twilight of the Habsburgs: The Life and Times of Emperor Francis Joseph,* Grove Press, New York, 1994.

Planck, Max, *The Origin and Development of the Quantum Theory* (Nobel Prize address, translated by H. T. Clarke and L. Silberstein), Clarendon, Oxford, 1922.

Planck, Max, *Scientific Autobiography,* Philosophical Library, New York, 1949 (translation of *Wissenschaftliche Selbstbiographie* and other essays).

Popper, Karl, *Unended Quest,* Open Court, Chicago, 1974.

Proust, Marcel, *In Search of Lost Time* (vol. 2), Modern Library, New York, 1992.

Pupin, Michael, *From Immigrant to Inventor,* Scribners, New York, 1924.

Roller, Duane H. D. (ed.), *Perspectives In the History of Science and Technology,* University of Oklahoma, Norman, 1971.

Rolt, L. T. C., *Victorian Engineering,* Penguin, London, 1970.

Russell, Bertrand, *History of Western Philosophy,* Allen & Unwin, London, 1946.

Schorske, Carl E., *Fin-de-Siècle Vienna: Politics and Culture,* Vintage, New York, 1981.

Schuster, Arthur, *Biographical Fragments,* Macmillan, London, 1932.

Segrè, Emilio, *From X-Rays to Quarks,* W. H. Freeman, San Francisco, 1980.

Sexl, Roman, and John Blackmore (eds.), *Ausgewählte Abhandlung der Internationale Tagung über Ludwig Boltzmann,* Akademische Druck- und Verlagsanstalt, Graz, Austria, 1981.

Sime, Ruth Lewin, *Lise Meitner: A Life in Physics,* University of California Press, Berkeley, 1996.

Stiller, Wolfgang, *Ludwig Boltzmann: Altmeister der klassischen Physik, Wegbereiter der Quantenphysik und Evolutionstheorie,* J. A. Barth, Leipzig, 1988.

Tennyson, Alfred Lord, "Lucretius," *Complete Works,* Houghton Mifflin, Boston, 1928.

Tolstoy, Ivan, *James Clerk Maxwell,* Canongate, Edinburgh, 1981.

Weaver, Jefferson H. (ed.), *The World of Physics,* Simon & Schuster, New York, 1987.

Wheeler, Lynde Phelps, *Josiah Willard Gibbs: The History of a Great Mind,* Yale University Press, New Haven, Conn., 1952.

Zweig, Stefan, *The World of Yesterday,* University of Nebraska Press, Lincoln, 1964.

Notes

In the following, *Leben* denotes Höflechner, volume 1; *Briefe* denotes Höflechner, volume 2; and *Meyer* denotes Stefan Meyer's reminiscence in Höflechner, volume 3. *Briefwechsel* denotes Flamm's compilation of the letters between Boltzmann and Henriette von Aigentler. *PopSchrift* denotes Boltzmann's *Populäre Schriften. LLP* denotes Blackmore (ed.), *Ludwig Boltzmann; His Later Life and Philosophy.* This collection includes translations of some of Boltzmann's own writings, extracts from some of his letters, and a selection of letters and reminiscences concerning Boltzmann.

Unless stated otherwise, translations from German are mine.

Introduction

vii *I don't believe that atoms exist:* Boltzmann reported these words of Mach's in his 1903 inaugural philosophy lecture (PopSchrift, p. 338) but does not say exactly when the incident occurred. Höflechner (Leben, p. 183) argues that the most likely date was January 1897, when Boltzmann delivered one of his first philosophical lectures, called "On the Question of the Objective Existence of Events in Inanimate Nature" (PopSchrift, p. 162). This was after Mach had returned to Vienna but before he suffered his stroke.

xi *ran around in my head:* PopSchrift, p. 338.

Chapter 1: A Letter from Bombay

1 *Nothing but nonsense* and subsequent extracts from reviewer's comments are from Rayleigh's introduction to the published version of Waterston's work in *Philosophical Transactions of the Royal Society*, vol. 5, 183, 1892, p. 1.

4 *clothes hung above a surf-swept shore:* Lucretius, book 1, l. 305.

4 *For surely the atoms did not hold council:* Lucretius, book 1, l. 1018.

6 *Yet often when the woman heard his foot:* Tennyson, "Lucretius," *Complete Works*, p. 275.

6 *eating, drinking, copulation, evacuation, and snoring:* Russell, footnote, p. 249; the remark is attributed to Epictetus.

7 *When the atoms are carried straight down:* Lucretius, book 2, l. 220.

9 *It seems probable to me:* from *Newton's Opticks*, quoted by Cercignani, p. 51.

11 smooth or jagged atoms; Lucretius, book 2, l. 400.

11 *By good luck, the atomists hit on a hypothesis:* Russell, p. 85.

19 *brought out considerable abuse:* from the entry for Waterston in the *Dictionary of Scientific Biography.*

20 *The omission to publish it* and *Perhaps . . . a young author:* Rayleigh's introduction to Waterston cited earlier.

Chapter 2: Invisible World

22 *the prototype of the impractical academic:* PopSchrift, p. 102.

22 *So, you would have done a better job than Beethoven:* PopSchrift, p. 238.

30 Boltzmann recalled later that he knew no English: PopSchrift, p. 96.

31 *olympian cheerfulness* and *It never occurred to me:* PopSchrift, p. 102.

32 *regional finance commissar (Finanzbezirkscommissar):* Leben, p. 1.

32 *stubby fingers and pudgy hands:* Meyer, p. 3.

34 The tiniest sliver of a clue: PopSchrift, p. 162. The lecture in which Boltzmann recounts this incident is the one delivered in January 1897, after which Mach is thought to have declared his disbelief in atoms.

34 *always serious:* Briefwechsel no. 7.

35 *the China of Europe:* Crankshaw, p. 11.

36 *Ah, but is he a patriot for me:* Crankshaw, p. 14.

36 *positive sciences had to be learned:* Barea, p. 132. The remark is by Friedrich von Gentz, an adviser to the Austrian politician and diplomat Metternich.

42 *I am very pleased by the work of your outstanding student:* PopSchrift, p. 100. The original letter from Maxwell does not survive. The wording given here is a reconstruction of Maxwell's English from Boltzmann's recollection in German.

Chapter 3: Dr. Boltzmann of Vienna

43 *thinner and somewhat older:* from Koenigsberger's autobiography, excerpted in Leben, p. 21.

44 *sweet, fat darling:* D. Flamm, in Cohen and Thirring, p. 9.
44 *Herr Professor, you have made a mistake:* Schuster, p. 221, who claims he heard from Koenigsberger that these words are exactly what Kirchhoff told him immediately after the incident.
45 *helpful and kindly:* PopSchrift, p. 53.
45 *he let me know he had already noticed it:* Briefe no. 3 and footnote.
46 *who within a short space of time:* Leben, p. 16.
47 a converted priest's residence: Leben p. 18.
49 the "Reichschancellor" of German physics: Cahan, p. 59.
50 *bowed before his master:* Pupin, p. 231.
50 *not so accessible:* Briefe no. 5.
50 *You are in Berlin now:* PopSchrift, p. 102.
52 *Elegance . . . is for the tailor:* this remark and variations on it are recorded by a number of people, including Franz Streintz, a Graz colleague of Boltzmann's, Leben, p. 64; Arnold Sommerfeld, *Wiener Chemiker-Zeitung,* February 1944, p. 25; and Clemens Shaefer, Stiller, p. 135.
56 *it is very important:* Leben, p. 45.
59 *there is something else close to my heart:* Briefwechsel no. 7.
59 *it's admittedly a presumptuous request:* Briefwechsel no. 8.
59 *a lasting sympathy between us:* Briefwechsel no. 15.
62 *is an advantage for our personal life:* Briefwechsel no. 33.
63 *so tremendously expensive:* Briefe no. 19.
64 *a strong, heavy-browed man:* Kienzl's recollection quoted by Stiller, p. 16.
65 *if he had to make a proposal:* Briefwechsel no. 94.
65 *industrious but no genius:* Briefwechsel no. 124.
65 *That all the beautiful hopes for our honeymoon:* Briefwechsel no. 123.
65 *There is really nothing more I can do in Vienna:* Briefwechsel no. 125.
66 so he bought a cow: Meyer, p. 3.
67 *well-organized institute:* Leben, p. 71.

68 *with the physics genius Ludwig Boltzmann:* Leben, p. 89.
68 *the prototypical unworldly scholar:* Stiller, p. 16.

Chapter 4: Irreversible Changes

69 *The true logic for this world:* letter from Maxwell to Campbell, Campbell and Garnett, p. 97.
71 *First the variations in velocity:* PopSchrift, p. 73.
71 *those learned Germans:* Knott, p. 115.
74 *style is the man:* comment by Clemens Schaefer, quoted by Stiller, p. 134.
77 *Concerning Demons:* Knott, p. 214.
77 *the German Icari:* Knott, p. 116.
78 *Oh! I'm so glad:* Tolstoy, p. 14.
78 *Show me how it doos:* Campbell and Garnett, p. 16.
78 floating to shore on his bagpipes: Campbell and Garnett, p. 3.
79 *My dear Mr Maxwell:* Campbell and Garnett, p. 39.
79 *Let Pedants seek for scraps of Greek:* Campbell and Garnett, p. 384.
80 *Kant's Kritik of Pure Reason in German:* Campbell and Garnett, p. 87.
80 *spasmodic:* Tolstoy, p. 76.
81 *Was it a God:* Broda, p. 33.
82 *his full share of misfortunes at the blackboard:* Tolstoy, p. 100.
83 *By the study of Boltzmann:* Knott, p. 114.

Chapter 5: "You Will Not Fit In"

91 allowed one equation: see Hawking's remark in his acknowledgments.
94 *I fear that my earlier principal was correct:* Briefe no. 146.
95 *He did not rest:* from Franz Streintz, who had been a physics student in Graz, Leben, p. 63.
98 *a highly astute, outstanding mathematician:* Leben, p. 99.
99 *if the pigs with the left-curled tails:* LLP, p. 202.
101 *would regret it most keenly:* Briefe no. 225.

102 *Herr Boltzmann, in Berlin you will not fit in:* from Ostwald's autobiography, quoted in Leben, p. 99.

102 *by accepting my new appointment:* Briefe no. 238.

103 *Since I have seen for myself:* Henriette's letter to Schulze is reproduced in Hörz and Laass, p. 111.

104 *poor friend would only attain peace:* Schulze's letter is reproduced in Leben, footnote, p. 111.

105 *in the greatest agitation:* Briefe no. 242.

106 *overstrained nerves:* Briefe no. 251.

106 *Many [people] who were not to be had:* PopSchrift, p. 408.

108 *When I learned, a few days ago:* PopSchrift, p. 76.

Chapter 6: The British Engagement

110 Charles Darwin himself may well have become a country vicar: Browne, chap. 5.

111 *nothing would be uncertain:* Laplace, quoted by Cercignani, p. 55; the full statement can be found in Weaver (ed.), vol. 1, p. 582.

111 *If the law of forces were known:* Boscovich, quoted by Cercignani, p. 55.

112 *for every configuration;* E. P. Culverwell, *Philosophical Magazine,* vol. 30, 1890, p. 95.

112 a longer analysis from Burbury: S. H. Burbury, *Philosophical Magazine,* vol. 30, 1890, p. 298.

113 *The part which Prof. Boltzmann took in these discussions:* G. H. Bryan, *Nature,* vol. 74, 1906, p. 569.

113 *an odd little fellow:* comments by H. Nagaoka are translated into English in LLP, p. 5.

114 *My mind absorbed avidly:* and subsequent remarks by Planck are from his *Scientific Autobiography.*

115 *the second law of the mechanical theory of heat:* Kuhn, p. 23.

115 *the remarkable physical insight and mathematical skill:* Kuhn, p. 22.

116 *is determined not by probability but by mechanics:* Kuhn, p. 27.

117 *Will some one say exactly:* Culverwell, *Nature*, vol. 50, 1894, p. 617.

117 *there is a general tendency:* Burbury, *Nature*, vol. 51, 1894, p. 78.

117 Boltzmann's reply to his English critics: *Nature,* vol. 51, 1895, p. 413.

120 *our lives have traveled so far apart:* Briefe no. 316.

120 *I have often noticed:* Briefe no. 320.

121 *his ideal:* Briefe no. 328.

121 *no one could say a word against me:* Briefe no. 338.

Chapter 7: "It's Easy to Mistake a Great Stupidity for a Great Discovery"

123 *would like to bring about:* Briefe no. 399.

125 *bungler:* Briefe no. 301.

125 *you can hardly know:* Briefe no. 309.

125 *against the dogma:* Briefe no. 305.

127 *I hope I can count among my closest friends:* PopSchrift, p. 105.

127 *things went hard:* translation of Helm's letter in LLP, p. 49.

128 *closed antagonism:* Ostwald's autobiography, quoted by E. N. Hiebert, in Roller (ed.), p. 68.

128 *the energeticists were thoroughly defeated:* Leben, p. 169.

128 *the fight of the bull with the lithe swordsman:* A. Sommerfeld, *Wiener Chemiker-Zeitung,* February 1944, p. 25.

129 *cheerful, friendly Graz:* Blackmore, *Ernst Mach,* p. 38.

129 Mach's name came up: Leben, p. 121.

130 *fortwürsteln:* Crankshaw, p. 271.

134 *sehr talentlos:* Blackmore, *Ernst Mach,* p. 9.

135 *school cleverness and slyness:* Blackmore, p. 11.

135 *Kaiser Franz had let the Austrian universities go to the dogs:* Blackmore, p. 13.

138 *the object of natural science:* Mach, quoted by Blackmore, p. 85.

138 *We say now that water consists of hydrogen and oxygen:* Mach, quoted by Blackmore, p. 86.

141 *Did I return to Vienna as the gravedigger:* PopSchrift, p. 93.
142 *Herr Zermelo's paper* and *is like a dice player:* Hasenöhrl, vol 3, pp. 568 and 576.
142 *the English kinetic theories:* Poincaré's comments, quoted by Cercignani, p. 100.
145 *whenever other occupations allow me:* Briefe no. 403.
145 *the last pillar* and subsequent quotes from A. Höfler, M. Smoluchowski, and G. Jaffé LLP, pp. 75–76.
145 *Now I come to a delicate point:* Briefe no. 427.
146 *Whether I will soon be alone:* Briefe no. 428.

Chapter 8: American Innovations

147 The first steamship to cross the Atlantic: details from Rolt, chap. 3.
149 *the monotonous working life of Vienna:* Leben, p. 187.
149 *rather boring* and subsequent quotes: Fasol-Boltzmann, pp. 27–39.
150 *Effusiveness was foreign to his nature:* Wheeler, p. 53.
152 *a most important American contribution* and subsequent quotes: Niven, vol. 2, p. 426.
155 *I myself had come to the conclusion:* Wheeler, p. 138.
155 *This work had the greatest influence on my development:* Ostwald, quoted by Wheeler, p. 100.
156 *In justifying his theorems:* from "On Energetics"; and *which he had discovered by a different method:* from "On the Development of the Methods of Theoretical Physics in Recent Times," both translated in McGuinness (ed.), pp. 39 and 98.
158 *the impossibility of an uncompensated decrease of entropy:* from Gibbs's 1876 paper, in Bumstead and van Name (eds.), p. 167.
159 *we avoid the gravest difficulties:* from Gibbs's *Elementary Principles in Statistical Mechanics,* quoted by Wheeler, p. 150.
160 *in many places it seems evident:* from Boltzmann's *Lectures on Gas Theory,* vol. 2, translated by Brush, p. 406.

161 *I have read some metaphysics:* Knott, p. 215.

163 *Just when I received your dear letter:* Briefe no. 462.

163 *I am conscious of being only an individual:* from the introduction to Boltzmann's *Lectures on Gas Theory,* vol. 2, translated by Brush.

163 *I feel like a monument of ancient scientific memories:* from "On The Development of the Methods of Theoretical Physics in Recent Times," translated by McGuinness, p. 82.

Chapter 9: The Shock of the New

167 *the law of causality is sufficiently characterized:* from Mach's *Conservation of Energy,* quoted by Blackmore, *Ernst Mach,* p. 86.

170 *a sort of well-arranged catalogue of facts:* quoted by L. Badash in Roller (ed.), p. 89.

170 *Every hypothesis must derive indubitable results:* Boltzmann in *Nature,* vol. 51, 1895, p. 413.

170 *we infer the existence of things:* from "The Second Law of Thermodynamics," translated by McGuinness, p. 16.

171 *there are no absolutes:* reconstructed from Boltzmann's philosophy notes by Fasol-Boltzmann, p. 84.

171 *Phenomenology believed that it could represent nature:* from "On the Development of the Methods of Theoretical Physics in Recent Times," translated by McGuinness, p. 97.

172 *irresistible analogy:* Mach's phrase and his dismissal of Boltzmann's response quoted by Blackmore, *Ernst Mach,* pp. 36 and 206.

173 *mathematics is economically arranged experience of counting:* Briefe no. 447.

173 *those who consider themselves opponents of Mach:* quoted by Blackmore, *Ernst Mach,* p. 141.

174 *would not associate himself with energeticism:* from Ostwald's autobiography, quoted by Blackmore, p. 118.

174 *it was simply impossible to be heard* and *a second to Boltzmann:* Planck, *Scientific Autobiography,* pp. 30 and 32.

175 *will be easier and more promising:* letter from Planck to Leo Grätz, quoted by Kuhn, p. 28.

177 *it gave me particular satisfaction:* Planck's Nobel Prize address.

177 *false prophets:* Planck, quoted by Blackmore, *Ernst Mach,* p. 220.

178 the precocious young Viennese poet Hugo von Hofmannsthal: see Janik and Toulmin, p. 113; and Barea, p. 295.

180 *you are not personally angry with me:* Briefe no. 483.

180 *In Munich he was very happy:* Leben, p. 204.

181 *I cannot conceal from you:* Briefe no. 540.

183 *a stranger in this world:* Ostwald, p. 404.

183 *the prospect of winning you:* Briefe no. 541.

183 *rarely have members of the philosophical faculty:* letter translated in LLP, p. 62.

184 *for me personally it would be a hard blow:* Briefe no. 543.

184 *I have never concealed:* Briefe no. 546.

184 *I hope you will be happy in Leipzig:* Briefe no. 547.

185 *almost pathological:* Hartel's memo to the emperor reproduced in Leben, p. 210.

186 *real psychosis:* Leben, p. 212.

186 *neurasthenics find it impossible:* Proust, p. 253.

186 *Dearest Mama:* Briefe no. 557.

Chapter: 10 Beethoven in Heaven

188 *Bald bin ich dort:* the entire poem is in Leben, p. 221.

189 *Then, when unexpected Sleep:* Campbell and Garnett, p. 387.

190 *despite the grateful reception:* Ostwald, p. 405.

190 *Anyone who got to know:* recollection by Theodor des Coudres, quoted by Stiller, p. 36.

191 *I really haven't been feeling well:* Briefe no. 569.

191 *Papa sweats and swears all the time:* Fasol-Boltzmann, introduction.

192 *almost pathological ambition:* Hartel's memo is reproduced in Leben, p. 234.

192 *must have felt that he was on the defensive:* LLP, p. 89.

192 reported suicide attempt in Leipzig: Briefwechsel, introduction.

193 *north German protestant way of life:* Briefe no. 611.

193 *somewhat nervous and confused:* Briefe no. 579.

193 Hartel asked the emperor: Schuster, p. 221.

194 *if a fire breaks out:* Sime, p. 11.

194 *His lectures were the most beautiful:* Meitner, quoted by Broda, p. 11.

197 *The little house we bought in Vienna:* Briefe no. 610.

197 *Dingler's journal* and following anecdotes: Meyer, pp. 6 and 7.

198 *Papa is neurasthenic:* and subsequent quotes, Fasol-Boltzmann, introduction.

198 *How do I come to be teaching philosophy?* LLP, p.106. (I have slightly altered the wording to improve the sound of the English.)

199 *what an unclear, senseless torrent of words:* PopSchrift, p. 341.

199 *Is there any sense at all:* Briefe no. 642.

200 *lively and animated:* Henriette's words are reproduced in Briefwechsel, introduction.

200 *provide a small moment of happiness:* LLP, p. 111.

200 born in the night between Shrove Tuesday and Ash Wednesday: Meyer, p. 8.

200 *powerful nervous depression:* Briefe no. 630.

200 *very inferior:* and subsequent excerpts Fasol-Boltzmann, introduction.

201 *The amazing thing:* Millikan, p. 85.

201 *Proof that Schopenhauer:* from "Über eine These Schopenhauers," in PopSchrift, p. 385.

202 *steps very rashly from one point to another:* Briefe no. 649.

203 *See that ship over there!* and subsequent extracts, from "Reise eines deutschen Professors ins Eldorado," in PopSchrift, p. 403. Full but awkward English translations are given by Cercignani and in LLP; a more fluent but

partial translation by Bertram Schwarzschild appeared in *Physics Today,* January 1992, p. 44.

204 *Dass ist the truth:* LLP, p. 203.

204 *I were extremely glad:* Briefe no. 323.

205 *somewhat deficient, to put it mildly:* quoted in Leben, p. 280.

Chapter 11: Annus Mirabilis, Annus Mortis

206 *Now there is only the trifling train journey:* from "Reise eines deutschen Professors ins Eldorado," in PopSchrift, p. 435.

207 *Boltzmann is quite magnificent:* LLP, p. 32.

212 *The observed motions of very small particles:* Hasenöhrl (ed.), p. 572.

214 *Before that Pestalutz does it, I will do it:* PopSchrift, p. 406.

214 *How I envy you your constant cheerfulness:* Briefe no. 684.

215 *Unfortunately, things are not going especially well:* Briefe no. 685.

215 *unhygienic:* Meyer, p. 8.

215 *heart-rending groans:* Ludwig Flamm (Boltzmann's son-in-law), *Wiener Chemiker-Zeitung,* February 1944, p. 28.

215 *I would not have believed I could come to such an end:* A. Höfler, quoted in LLP, p. 208.

215 *no pressing pedagogical need will be damaged by that:* Leben, p. 287.

215 Description and quotes concerning Boltzmann's death are from the *Neue Freie Presse* of Vienna for Sept. 7 and Sept. 8, 1906.

217 For the rest of her life: D. Flamm (her son), in Briefwechsel, introduction.

219 *rides Mach's poor horse to exhaustion:* Blackmore, *Ernst Mach,* p. 256.

219 *Mach was a good experimental physicist:* Blackmore, p. 258.

222 *Nothing more will come from me:* Meyer, p. 8.

223 Suicide was pervasive in Vienna: Morton, p. 133.

223 An architect . . . killed himself: Barea, p. 240.

223 One of Franz-Josef's finance ministers had killed himself: Crankshaw, p. 173.

224 *He may have been wounded:* Sime, p. 15.
224 *those who strive* and subsequent exerts from obituaries for Stefan and Loschmidt, in PopSchrift, pp. 92 and 228
225 *The overall struggle for existence:* PopSchrift, p. 40.
226 *We see how it was useful and important* and *We understand why only those individuals:* PopSchrift p. 314.

Postscript

230 Wittgenstein . . . had hoped to study under Boltzmann: Malcolm, p. 3.
230 Schrödinger . . . enrolled as an undergraduate: Moore, p. 39.
231 Boltzmann's coffin was removed: Meyer, p. 8.

Index